THE GREAT
SCHOOL LEGEND

THE
GREAT
SCHOOL
LEGEND

*A Revisionist Interpretation
of American Public Education*

COLIN GREER

BASIC BOOKS, INC., PUBLISHERS

New York   London

FOR SIMON

Foreword

Colin Greer has written an important book which deserves to have a far-reaching impact on educational thought in America: he has dared to question the accuracy of one of the most sacred —and durable—legends in American education.

The legend has it, that once upon a time the public school was an effective antipoverty agency, that took poor immigrant children and taught them so well that eventually they became affluent Americans. The reality, as Colin Greer shows in this book, was quite different: the public schools of the late nineteenth and early twentieth centuries did not help poor children, but instead, failed them in large numbers and forced them out of school. Indeed, the actual function of the public school was just the reverse of the legendary function; it certified the children of the poor as socially inferior at an early age, and thus initiated the process that made many of them economically inferior in adulthood and kept them poor. In current terminology, the school was an agency of negative credentialism.

Although this book is historical, it has many lessons for the present. Because the legend of the school continues to exist, the social policies built on the legend also continue to exist; and even now there is faith that if only a magic formula could be found, today's schools would enable today's poor to become members of the affluent society. And because education is still thought to be a vehicle for moving people out of poverty, the

search for other, more effective vehicles is neglected. As Colin
Greer puts it in his Introduction:

So the legend supports a social policy comfortable in the notion
that the agency for the amelioration of most social problems already exists—and that those problems whose solutions now elude us
will be resolved shortly, or are beyond solution, through no fault
of that great nation, but because of deficiencies in particular people
who cannot seem to solve their problems as countless other Americans have before them.

Or to put the matter more bluntly, we blame the blacks,
Puerto Ricans, Chicanos and others for being unable to use
the school as we imagine the poor European immigrants did
before them. And by blaming the victims, as William Ryan has
pointed out, we relieve ourselves from the responsibility of doing
anything about poverty.

But *The Great School Legend* has an even more tragic impli-
cation for today. Although the schools Greer studied failed their
poor students, many of them were able to enter the economy
because of the availability of unskilled jobs that required no
education. Since then, the economy has changed drastically, and
almost all available jobs require some education and more of it
all the time. Consequently, the school now *has* to learn to teach
poor children or else those children will have less chance than
previous generations had to escape poverty.

While Colin Greer's book is a study in educational history,
it also debunks some noneducational legends. He shows that the
economic success of the European immigrants has been highly
overrated, both in the past and in the present, since many of
the original immigrants remained poor all their lives, and raised
children and grandchildren who also remained poor. As he
points out, the descendants of Italian and Polish immigrants,
among others, are only now going to college in respectable numbers, three, four, or five generations after their ancestors came
to America.

The findings in *The Great School Legend* raise additional
doubts about the currently popular belief that blacks are only

the latest ethnic groups to come to the cities to find their fortune: and thus being no different than the European immigrants, will in time make it economically and socially in much the same way. Instead, his data indicate that at least since the nineteenth century, urban black school children have always been treated separately from and more unequally than poor whites, and that this treatment has changed little over the decades, although it is now perhaps more humanitarian in intent if not in outcome. Mr. Greer's study thus provides additional evidence for the theory that from the start blacks were not given a chance to enter the economic mainstream of the cities, and that something other than the laissez-faire approach by which the immigrants were integrated into that mainstream will be required.

One of the most fascinating questions raised here is why "our official historians have mistaken the rhetoric of good intentions [of the public school] for historical reality," and why so many continue to believe the rhetoric and ignore the reality. As a sociologist, I am persuaded that such historians are wrong simply because, given the acknowledged failure of today's schools to teach poor children, there is no reason to believe that they were as radically different in the past as the legend claims. Institutions do not change drastically over time unless other parts of society also change, and there have been no drastic changes in the position of the poor in American society during the last century (whatever the reduction in their numbers), in the role of the school in the community and the class structure, or in the kinds of people who run and teach in the schools.

Still, there are understandable reasons why school history has emphasized rhetoric over reality. To begin with, much of it is "in-house," as Colin Greer indicates; written by educators, themselves deeply dedicated to the public school, or by other academics who are, after all, also in the school business, and thus not always able to see their own institution with detachment. Furthermore, there are data which suggest that the historic public schools had some success, since the minority of children who graduated from them did well in later life. It is therefore possible that the school was responsible. On the other hand, it is also possible that these children succeeded as adults because they came from affluent homes, and that the school had little to do with their achievement. Yet even if the school could be shown to have helped youngsters from affluent homes, there is no justification for assuming that it helped the majority who were poor, particularly since they became drop-outs or push-outs long before graduation.

In addition, one can point to at least one ethnic group which achieved some success through the school, the Jews, although I have long been doubtful that the school deserves as much credit for the upward mobility of poor Jewish children as it has received. A small number of well-publicized successful business-men and famous intellectuals of poor origin were aided in their rise by the school, and so were some and perhaps even many other Jewish children who entered the lower middle class. Still, many others escaped from poverty without help from the schools, and yet others never got out. But even if the Jewish success story could be entirely attributed to the public school, historians, Jewish and otherwise, should have been aware of the great diversity among the immigrants, and the diversity in school experiences it produced. For example, they should have realized that Sicilian farm laborers and Polish peasants, to mention just two groups, came to America with far more meager economic and social resources than did the Jews, who were already urbanized in Europe and could therefore more often establish themselves as artisans, peddlers and petty retailers. Not only did many Jews have enough economic security to send their children to school in America, but unlike most of the other immigrant groups, they had themselves attended school in Europe and could thus give their children more of the intellectual and emotional support they needed to benefit from schooling.

Finally, while the documentary evidence with which historians work contains endless rhetorical statements about school performance, the actual performance of the school was—and continues to be—difficult to study; for like all institutions for children, schools are almost impossible for adult researchers to observe directly. What actually happens in the classroom must therefore be learned from other adults, particularly teachers and administrators, and they have little incentive to admit their inability to achieve their own goals. And what is true of today's schools was even more true in the past, when disenchanted teachers like Herbert Kohl and Jonathan Kozol did not write their memoirs.

Nevertheless, many of the statistical data Mr. Greer presents have long been available to scholars, so that one must suspect that they have been ignored because of the deep faith in the educational legend. Why this faith is so strong is itself a puzzle, for as the Coleman Report indicates, school has a less pervasive influence on students than home background, and besides, if schools were as potent as the legend proposes, why has American culture until recently portrayed teachers as ineffective old maids?

Perhaps the legend has survived so long because schools have been perceived as altruistic institutions above politics, untouched by the self-interest that pervades the economy, and detached from the social structures that perpetuate poverty. As a result, until the 1960s, it had not even been thought necessary to ask whether school performance actually coincided with the rhetoric. Moreover, the realization that the school does not teach poor children and that it is a middle-class institution which itself contributes to the persistence of poverty is fairly new, and this realization has not been around long enough to put an end to the legend.

The legend has also endured because in the old days, when unskilled jobs were still plentiful, many of the immigrant children who were failed by the schools entered the economy by the strength of their muscles. And because the death rate among the poor was high, some of those who continued to fail after school did not survive and, leaving no trace, could easily be forgotten. Today, only the successful immigrants are remembered, which is why we can indulge in a romantic view of the old Lower East Side and other historic slums, ignoring that they were terrible places, probably more terrible than they are today.

The time may now be ripe to inter the great school legend and to find new ways of making the school into an effective antipoverty agency. Consequently, Colin Greer—who is himself "in-house" enough to retain a fragile faith in the school—ends his book with a dramatic appeal for a radical strategy that would force the school to live up to the very legend he has so success-fully debunked. In calling for "maximizing the tension between school rhetoric and reality," a strategy similar to that proposed by Richard Cloward and Frances Piven for transforming the public welfare system, he suggests that the school might yet become an "agent for major change in the society." His faith is tempered by realism, however, for he also suggests that "there is no precedent . . . to comfort the belief that there is hope for society through schools."

Educators like himself must have a strategy if they are to continue working in the schools, and his strategy is certainly worth trying, though there is little cause for optimism. Recent studies of contemporary public school systems have shown over and over again that even if the majority of their students are poor, control of these systems is firmly in the hands of the middle-class teachers, administrators and interest groups, white and black, some of whom have themselves only recently escaped from poverty. Whether they are old or new middle class, how-ever, they often believe that those who have not escaped from poverty are unteachable; and like many other middle-class Americans, they are not entirely sure that the poor ought to be allowed to enter the economic mainstream and thus compete with them and their own children for jobs. This is not to deny that poor youngsters are hard to teach, but even if they were not, the professionals and political groups which control the schools have little incentive to translate the great school legend into reality.

Moreover, one of the most important implications which can be drawn from Colin Greer's study is that in the past, poor people did not succeed economically through the school, but that as they succeeded economically, they could exert pressure on their children—and the teachers—to make sure that their children would succeed in school. This is also one of the implications of the Coleman Report, and suggests that educational success follows upon economic success, not the other way around. Consequently, unless the schools are able to reverse their historic role, which I doubt, the only way today's poor can escape is through the economy, and only then will they be able to see to it that their children can succeed in school.

Educators like Colin Greer need an educational strategy against poverty, but for those of us who are concerned with an overall strategy, his book reconfirms the belief that such a strategy must be mainly economic. Poverty is likely to be eliminated only if the economy and the federal government jointly create decent, secure and well-paying jobs for today's unemployed and underemployed, buttressed by whatever on-the-job training is necessary and backed up by adequate income grants for adults who can not work. Only when such a program has created a generation of economically secure graduates from poverty will it be possible for the public school to do the job that the great school legend so wrongly claims it once did.

HERBERT J. GANS

Acknowledgments

I want to thank Bonnie Prives for typing my obstinate hen-scratching and then accepting revisions as if they were both inevitable and no trouble at all. I am greatly indebted to Erwin Glikes, my editor, who was in the best sense of the word a teacher—helpful in clarifying confused thoughts but respectful always that it was my thing we were working on. I am deeply grateful to my wife Valerie who believed it was important to write this book even if it meant more than her fair share of kitchen duties and what must have been long, quiet evenings. Finally, I want to thank Simon, my marvelous little boy, who has taught me to really value childhood, to trust its creative energy, and to want to defend children against their many powerful enemies.

Contents

PART

I

The Legend

PART

II

What Really Happened

THE GREAT
SCHOOL LEGEND

Introduction

Once upon a time there was a great nation which became great because of its public schools. That is the American school legend. And the faith in that legend is so great that most social problems—the bumps and jolts which interfere with the great nation's running smoothly—are seen as unique phenomena, exceptions to the rule, phenomena which the schools will eventually mitigate. So the legend supports a social policy which is secure in its faith that the agency for the amelioration of most social problems already exists—and that those problems whose solutions elude us now either will be resolved or are beyond solution, through no fault of that great nation but because of deficiencies in particular people who cannot seem to solve their problems as countless other Americans have before them. It is a pernicious legend, then, because it justifies the exclusion of millions who will never share in America's greatness as long as the legend persists.

Every schoolchild and certainly every education major learns the same heart-warming story about the history of our public schools. The public school system, it is generally claimed, built American democracy. It took the backward poor, the ragged, ill-prepared ethnic minorities who crowded into the cities, educated and Americanized them, and molded them into the homogeneous productive middle class that is America's strength and pride. But that story is simply not true. Worse yet, the "Great School Legend" is largely responsible for today's schools' resistance to needed change. Having examined the records of a number of major urban school systems, I will attempt to report here the facts obscured by this legend, and to suggest the consequences of our misguided faith in the schools. The rate of school failure among the urban poor, in fact, has been consistently and remarkably high since before 1900. The truth is that the immigrant children dropped out in great numbers—to fall back on the customs and skills their families brought with them to America. It was in spite of, and *not* because of, compulsory public education that some eventually made their way.

But the unchallenged persistence of the school legend has the most serious and harmful consequences for today's—largely black—urban poor. They are held responsible for failing to make the same good use of the schools their predecessors did. In assumptions and practice, the urban schools remain essentially unchanged—and the poor continue to fail. Our official historians have mistaken the rhetoric of good intentions for historical reality. I hope to offer here some initial attempt at sorting out the immense differences between them.

The American public school has a sacrosanct place in the democratic rhetoric of the nation. The school's development and expansion are believed to be an essential part of a continuing national response to the poor and the less than fortunate at any given time in the face of rapidly changing economic and social conditions. Layers of education—additional years—have been added until free public education has been expanded from a few years of grade school to a college education for all. We have added new wings and built a high-rise structure on top of the little red schoolhouse to contain more and more students for longer and longer periods of time, tacitly assuming the success of the existing structures in planning our new additions. We pile more of the same on top of what we have. When we "reevaluate," we change a few things to keep the rest pretty much the same.

Every time we add a new level we contribute to the mistaken

belief that schooling beneath that level must be working well since, after all, it seems to be successfully preparing students for more advanced levels of work. This is hardly the case. Often all we do by adding new layers on top of the existing structure is to stretch out the process of failure.

Certainly it cannot be said that the public schools have been unwilling to accept more students from among previously excluded groups in American society. We even recognize that children who don't do well in school suffer from the fact that our measuring instruments are culture-bound, but the lesson we draw is that the children need help to do better. They may. Yet the problem is not with the test, but in the schools. The schools are more culture-bound than any of our most vociferously denounced tests, culture-bound not only in the sense that they present a different and more threatening set of cultural norms than those many of the children have experienced, but in the sense that the schools projects and imposes deadening conventions. They translate spontaneity and impulsive creativity into the feelings of immobility, impotence, and anguish which dominate the lives of so many Americans of all ages. Our schools teach the compromises, the willing suspension of hope, which lead us as adults to conform to life in a society in which "marriage is a convention . . . of sexual attitudes . . . war is the convention of murder, plundering, and arson, diplomacy is a conventionalization of cheating and lying."[1]

I have little sympathy with a mechanical conspiratorial theory of history to explain the difference between the rhetoric expressing the American school dream and the empirical social reality which it obscures. I do not see men plotting devices behind closed doors to ensure profit—at least not often. And yet there are patterns in the reality which can only be explained as a result of a consistent triumph of the self-interest of those who "have" over the aspirations of those who "have not".

This book is an attempt to bring to the history of education the revisionist insight which has been influencing and reshaping so many of our traditional interpretations of other aspects of American history.[2] My aim is to urge us to reexamine the historical analysis of American education with an acute awareness of contemporary social crises.

The pain of more than 30 million desperately poor and hopeless Americans argues that we are in need of much more radical rethinking of our history and our present situation than we have been willing to undertake in the past. Our inability to make any substantial progress in ameliorating our most pressing social problems indicates the probability that our present analysis is at fault. Before we can effectively reformulate our policies we must test our historical and sociological theories against the facts. The time-honored faith that public schooling effectively paved the way to future mobility and status for generations of Americans has attracted devout followers undaunted by the schools' strangely inexplicable, disastrous inability to repeat the magic for today's poor. A legend has emerged and reigns triumphant: that American consensus out of diversity and material well-being after poor beginnings are miracles that have been accomplished with the public school acting as prime agent in the process. But we have yet to examine the importance of the fact that it was in the public school that this nation's long-standing fear of the foreigner, combined with the discovery of his sudden economic usefulness and uneasily, reluctantly refurbished a traditional American faith in the social power of free, public education.

Despite the availability of research which suggests that traditional optimism and self-satisfaction about the history of American education might be misplaced, the legend has gone largely unchallenged. Indeed, the story of the development of American public schools has consistently been told as the inevitable triumph of optimistic, democratic humanitarianism. What Samuel Eliot Morrison has called the "neo-liberal stereotype," dominant in American historiography since the conflict theories of Vernon Parrington, Frederick Jackson Turner, and Charles A. Beard, has been perpetually reinforced in more specifically educational historiography.[3] As a result, social problems in the school and in society look as though they are either unique in nature (the problem of the particular people concerned) or part of the familiar positive process—just another transitional stage in achieving the perennial consensus and continuity of "progress" which dominates the American past. All the evidence to support a contrary interpretation notwithstanding, it has been possible to tell this essentially conservative, static story in standard educational histories by citing, over and over, the rhetoric of intentions expressed by school idealogues and strategists in the past rather than by examining the facts of school life and the schools' effect on the people in them.

It is instructive to note how minimal the influence of some important research has been in weakening the legend. At times one suspects the researchers themselves may not have realized the full implications of their own findings.

Some recent examples of this include the work of David Tyack on how the public schools stressed "indoctrination" more than thinking in planning programs for the new immigrants and Tyack's documentation of how difficult it was for minority groups to override exclusionary school practices.[4] Maxine Greene has illustrated how secondary the concern with human problems was among schoolmen.[5] Raymond Callahan has shown the dominance of businessmen over schoolmen and school programs.[6] Charles Burgess and Charles Strickland have written persuasively about the elitism and racism of G. Stanley Hall, the pioneer in applying psychology to public education.[7] So too, the growing body of work on the long-standing deprived status of blacks has had too little impact on the established, historical reconstruction of public school development.[8]

Several of the very people whose studies have generated these potential challenges to conventional views manage to come away from their own work with strong traditional optimism. Tyack, for example, takes a benign view of attempts at educational reform, conceding the need to reform previous reforms for the sake of new, immediately excluded groups, thereby tacitly assuming that some people will always be excluded, but that their turn to be included will come—eventually.[9]

True, Callahan has shown how school people came under the sway and dominance of business and industrial groups after

1900. But he too somehow manages to look back to a time when he imagines schoolmen were free of this kind of dominance, for, understandably enough, he finds few examples of twentieth-century corporate pressures in the nineteenth century. He does not seek, and so does not find, the nineteenth-century parallels to modern business dominance.[10] Peculiarly enough, the school's subservience to society at a strikingly important period in modern history has been described in such a way as to leave the school itself a victim rather than an agent of the society which emerged.

My main concern in this book is to use the insights of recent research and my own work of the past few years to reevaluate the public school's role in the development of American society. Michael Katz, for example, has shown how the middle class, not the working man, dominated the nineteenth-century campaign for the establishment of common schools;[11] James Weinstein, Gabriel Kolko, and Roy Lubove have shown how "progressives," involved in the second major revamping of public education at the turn of the twentieth century, were overwhelmingly concerned about social order, anti-socialism, and the establishment of professional social welfare rules which placed insuperable obstacles between the spontaneous will to serve and those in need of help;[12] and my own work and the work of David Cohen suggest that the popular idea that immigrant social mobility came as a result of school success—contrasted with the blacks' unique academic and consequent economic failure—is quite illusory—and self-serving.[13]

In chapter 1, I report the traditional story, the legend enshrined in the standard historical analyses of public schools. Chapter 2 suggests how the schools and cities would look if, in fact, the "Rosy Picture" was accurate, and compares this utopia with how they do look today. In chapter 3, I analyze how the leading educational historians first established and continue to retail and reinforce the legend. In the fourth chapter, I begin to examine what really happened, starting with the conservative background of our first tax-supported schools. Linking the mid-nineteenth-century Common School Movement with the Abolition Movement, I have tried to emphasize the huge gap between rhetoric and reality in key American social reform movements. Chapter 5 attempts to suggest how little the schools really had to do with how those immigrants who did make it managed to do so. Finally, chapters 6 and 7 explore the reality of school performance for immigrants and blacks, relating school failure to the economic conditions and expectations of the wider society.

I do not expect that this book alone will explode so old and so strong a faith in the "Great School Legend." But I do hope that I will have made it wobble a little, expose its shaky foundations, and that this study will encourage others to do the work that remains to be done in order to dismantle it.

PART
I

The
Legend

I

The Rosy Picture

Let me attempt a fair, if compressed, summary of the "Great School Legend." As it unfolds, I do encourage your skepticism.

From the very beginning, the narrative runs, Americans depended on their schools. As Henry Perkinson summarizes it: "Alone in the savage wilderness of their new settlements, the earliest colonists had to rely upon schools and schoolteachers far more than they did in Europe."[1] As Arthur Lean put it in a review of the Educational Policies Commission 1955 report,

like the democracy of which they are a manifestation, public schools have justified the faith of the American people. Like other institutions, they are not perfect; like any institution, they have shortcomings. But their contributions have been significant and lasting. The United States would not be so democratic, so prosperous, so satisfying to the individual, and so strong in mind and spirit as it is today were it not for the nation's record in developing and supporting public schools.[2]

The report itself, sponsored by the National Education Association and the American Association of School Administrators and largely written by Lawrence Cremin of Teachers College, Columbia University, declared that "A source of profound strength lies in the American educational heritage . . . designed especially for their task, public schools have stood—and now stand—as great wellsprings of freedom, equality, and self-government."[3]

Frontier conditions made it impossible, so the story goes, for traditional parental responsibilities to be fulfilled without resort to a new supportive social agency. Increasingly, the school was expected to replace the family as the institution which would guarantee social stability and underwrite the national promise of self-improvement.

The new society was at once both eager and afraid of breaking out of old traditions. The wilderness required courage, spontaneity, and defiance. Children had to respect their parents, but they had to be ready to explore geographic and social ground beyond the parental community. Cultural dislocations—the culture shock, as Handlin and, for the schools, Bernard Bailyn, have termed it—led to Massachusetts' 1664 compulsory education law, which required formal schooling of all the Colony's children. An important formal amendment to the tradition of parental supervision and preparation was thereby introduced to the Colonies. Five years later Massachusetts adopted a law that required each town to provide schools and schoolmasters to fulfill the intent of the earlier legislation. An official state-designated, later state-run, structure was made responsible for seeing to it that there would be schools for the children. And so a very young America produced the first compulsory education laws of modern times.[4] In his influential interpretation of American education, Bernard Bailyn claims that the challenges of life in the wilderness combined with the new schools founded in that setting to encourage individuality and independence in the children. This can certainly be debated. But another assertion he makes in this connection cannot be argued: "the transformation of education that took place in the colonial period was irreversible. We live with its consequences still."[5]

Quickly, the legend continues, the very newness of the new nation being formed dictated other related purposes for the schools which, by the end of the seventeenth century, had been established by law in most of New England, the Middle Atlantic Colonies, and Pennsylvania. The American Colonies needed manpower from Europe. In addition to unskilled labor, the Colonies needed trained workmen with a large variety of skills to perform many different tasks—skills which in Europe were the preserve of guild systems with long apprenticeship routes to membership. The Colonies had no time for such rigidities, Daniel Boorstin tells us, nor indeed did they have the institutions to establish them in the New World. In Boorstin's view the newcomers not only had to adapt to an unfamiliar setting, but needed new institutions to do so. Above all, they organized institutions of learning—private colleges, and public schools. The uniformity of the language, the base on which Boorstin believes a single nation was early created in America, "depended on schooling and universal literacy."[6]

In his study of the middle schools (secondary education) during the eighteenth century, Robert Middlekauff traced continuous lines from the Puritan tradition of learning—the attempt to find practical applications of the Bible—to Benjamin Franklin's new secular definition of practical learning. Any one man might have to do anything during a lifetime in a frontier society. So, clearly, every child must be prepared to cope with the unexpected. This was the goal to which Franklin's prototypical "Idea of the English School . . .," presented to the Trustees of the Philadelphia Academy, was directed. It was a proposal for a permanent school from which boys "will come out . . . fitted for learning any Business, Calling, or Profession."[7]

Just as with the business of earning a living, so with the business of conducting a government of a republican nature. Once political ties with Great Britain had been severed, schools were quickly associated with the major task of educating intelligent citizens of the new republic. Such educational goals would, in the opinion of men like Thomas Jefferson, "illuminate, as far as practicable, the minds of the people at large" and thereby avoid deterioration of democratic government into a tyranny of one or of a few.[8] At least since the extensive establishment of common schools in the Jacksonian era, Rush Welter points out, Americans have "insisted upon the enlightenment of the people for expansive rather than restrictive reasons." The preservation of social stability was a national partner to "secure as yet unachieved ends."[9]

The traditional histories claim with evident pride that the new national government put itself solidly behind these goals. As evidence, they cite the Northwest Ordinance of 1787, which required each township in the Northwest Territory to set aside a mile square section of land for school purposes. During the early period of the Republic, each state government encouraged the establishment of schools. America was to be a free society where positions of power were open to all men of merit. Schools would ensure that men of talent might rise whatever their social and economic origins—"a national aristocracy of talent," Jefferson called it, moving from school to university and into positions of national leadership. "The first half century of our Republic," Ellwood Cubberley wrote at the turn of the twentieth century while new immigrants by the millions tested American democracy once again, "from an educational point of view, was largely given over to the principle that 'the whole state is interested in the education of the children of the state.' " Cubberley sketches a less harmonious picture than Middlekauff, but he too posits the existence of a national consensus about the goals of public education. Once the "dangerous and undemocratic" charity schools were superseded, Cubberley held, and the principle firmly established that a free common school education paid for by a general tax was "the common birth right of every American child," then the public school became "established permanently in American public policy." Between the Civil War and the period of heavy immigration during which he wrote, the public school system was "the prime essential to good democratic government and national progress."[10]

By the 1820s Jacksonian Democrats began to emphasize the greater importance of the public school as opposed to the private university as a source of leadership in a truly democratic society. The elitist, aristocratic remnants of European culture were challenged and substantially reformed through the agency of the "common" public school. The school's task now was "equalizing" rather than "selecting out," and the work was done to make the schools accessible to all, thereby guaranteeing equal access to opportunity. The schools were to be an agency for eliminating all privilege and destroying all elites by giving all men the same good "common" education: "common" as in common to all, not as in lowest common denominator.[11]

By the 1850s almost every Northern state had provided common schools for its citizens, governed and supported by the general tax pioneered in seventeenth-century Massachusetts. As Bailyn sees it, the fact that "American public schools had adhered to the goals of their founders, was to a considerable extent the result of the forms of institutional financing and control that emerged in the seventeenth century." Fifteen years later, the Southern states too began to establish statewide systems of public education. Alone with widespread public access to schools came the professionalizing of education and eventually pupil evaluations based on tests of intelligence. Responsible, well-trained school people would oversee the vital educational process and do their work by a scientific method; selection and encouragement on the basis of merit would replace all class privilege.[12] That was the promise.

The schools, then, were the institution through which the United States sought to hold itself together and prepare its citizens for the greatest of all experiments in democracy. In summing up the continuity in the thinking and public declarations of school statesmen from Horace Mann to William Torrey Harris (1840 to 1900), Lawrence Cremin writes, "common schools increased opportunity; they taught morality and citizenship; they encouraged a talented leadership; they maintained social mobility; they promoted social responsiveness to social conditions."[13] Harris, Federal Commissioner of Education during the height of American urbanization, industrialization, and immigration at the end of the nineteenth century, differed from Mann, Cremin points out, to the extent only that "Mann's Common School was to contribute substantially to fashioning an emerging social order governed by a new public philosophy of rational, humanitarian concern for the well-being and egalitarian opportunity afforded citizens in a framework of social stability; Harris was thereby to play a part in confirming an order that had already come into existence."[14] Or as Martin Dworkin put it, writing on John Dewey, "Progressivism, the creed of early twentieth century school reformers, was the democratic faith in the instrument of the common school, or public school, inherited from Jefferson and Mann—but now applied to the problems of training the urban and rural citizenry for industrial and agricultural vocations, and of acculturating or Americanizing the sweltering masses of immigrants."[15]

The end of the Civil War marked a major shift for the nation, the shift from farm to city. More immigrants came from Europe; they came in greater numbers than ever and settled more and more frequently in cities. All of the problems of old—fear for civilization, the need to meet the unexpected, the promise of egalitarianism—were to be tested in new terrain. Once again the nation looked to its schools.

Urbanization, once again, meant the absence of familiar institutions, it meant more diverse newcomers than ever before, it meant high rates of personal deterioration in unique geographic density, evidenced in high mortality rates, suicide, alcoholism, vagrancy, pauperism. Yet the school, with the assurance of its past successes, was to oversee the shift and translate the loss of traditional patterns into the opportunity to successfully tread new ground. In effect, the school's job was stabilizing a nation, respecting the integrity of ethnic group structures and identity, while at the same time freeing the individual to explore and advance in society regardless of his class or religion. It encouraged individual freedom from tradition, Nathan Glazer suggests, but did not demand it.[16]

As Perkinson sums it up, "The city child, especially the child of the newcomers, had generated both compassion and fear . . . He was in need of help. But he was also a threat, a threat to the working man, a threat to social customs . . . a threat to the future of American democracy. Partly from fear and partly from compassion, thirty-one states enacted some form of compulsory education law by 1900." By 1918 all the then forty-eight states had compulsory education laws.[17] The newcomers had to be contained and taught the ways of their new country.

The number of children attending school skyrocketed, requiring new organization, new teacher preparation, new curricula. This "transformation," as Cremin views it, of the public school in the city—the opening up of schools and therefore of American opportunity to millions of foreign-born and their offspring—is, in the analysis of most historians of education, evidence of and testament to a continuing success story and is the key to understanding the modern public school: it is by nature urban, bent on academic performance, with a virtual monopoly on career opportunity, and attended by many "newcomers." After all, thanks to the confidence the school inspired and the sensitive effectiveness with which it did its work, the floodtide of foreigners coming to America before World War I was accommodated, assimilated, and set on the road to the success they currently enjoy in the suburbs and more privileged parts of the city.[18] The latest floodtide of mass migration has been the relatively recent large-scale movement of Negroes to the cities. It is their situation which dominates contemporary school reports and most current accounts and statistics of academic failure. These new urban blacks are, in the optimistic description of Oscar Handlin and Iverne Dowie, the "new immigrants" and they are, Handlin asserts, merely following the patterns of struggle of previous minority groups. They are "repeating the experiences of earlier groups drawn into the processes of American democracy." For Handlin, like most social and educational historians, shares the faith he himself ascribed to the immigrants in his book, *The Uprooted*: "there is vaguely an understanding that the school will help them get on."[19]

So the school was established in the American imagination as the primary means whereby the nation could resolve most of its internal social conflicts, even those generated by diverse foreign immigration. "Give me your tired, your poor, your huddled masses yearning to breathe free, the wretched refuse of your teeming shore. Send these, the homeless tempest-tost to me, I lift my lamp beside the golden door!" was the faith which gave the Statue of Liberty life. And, according to the familiar narratives, the school eventually did make loyal Americans, productive workers, and affluent citizens out of all this human garbage. The tattered old-world cultures were transformed, giving fresh vitality to a new world. And the "immigrant problem" went away—thanks to the public school system.[20]

The same vague faith in the schools which Handlin alludes to—and shares—pervades most professional and popular analyses of current urban problems and of the school's role in ameliorating them. Poor blacks and other American poor—in the cities particularly—can look forward to the school fulfilling its historic mission, and in that tradition reforming itself markedly in the process. Concerned about reading scores, cognitive development, cultural deprivation, American sociologists and psychologists work against a background which, as the U.S. Riot Commission Report of 1968 puts it, portrays the children and grandchildren of immigrants moving up "to skilled, white collar, and professional employment." Miriam Goldberg, a psychologist noted for her research into problems of current school failure, more cautious than many of her colleagues, nevertheless accepts the story of the immigrants' "move from the status of alien slum dwellers to that of middle-class citizens in the course of one or two generations" as a logical frame of reference for addressing the current misery of urban poverty and exhibits an optimistic readiness to make the right kind of changes in order to achieve social progress through the school yet again. Bitterly aware of the schools' current failure to accomplish this end with poor children, sociologists and psychologists, black and white, have generally agreed that this is a new, contemporary phenomenon. The schools have done the job before, they say, they can do it again, though they may need some timely reform.[21]

Fred Hechinger of *The New York Times* recognizes such school reform takes time, that it "has always been a matter of island-building rather than a dramatic forward surging," but he believes gradualism has guarded against "headlong jumps into abortive novelties." As Hechinger sees it, after all, what we are waiting for and working toward in our current debate concerning experimentation with schools and the continued poor performance of urban poor people is reform enough to "recapture the melting pot function."[22] "Our public schools have, by any reasonable standard," Albert Shanker says, "enjoyed great success . . . Masses of immigrants, the poor, the illiterate have been educated and, through education, have achieved unprecedented upward social mobility." Black social scientists and black journalists—Kenneth Clark and Ernest Fergusson, for example —even black radicals like Eldridge Cleaver, Harold Cruse, LeRoy Jones, have each in their time asked for the chance previous minority groups had through schools—if not the melting pot function, at least the mobility function.[23] Recapture the openness of American society, they say, through equal access to schools and ethnic solidarity—the trusty weapons of yore. Assuming the past efficacy and the present potential of that openness, Daniel Patrick Moynihan looks to the school both to safeguard openness and to provide skills required for mobility. "Education in the United States," Moynihan tells us, "has had this deep social, and if you will political, purpose for well over a century now, and there is no sign of any diminishment in that intent."[24] For Edward Banfield, schools are already catching up with new urban social needs and are doing a fine job for those, white and black, ready to take advantage of them.[25] Paul Goodman, too, much less sanguine about current school directions, is still inspired by his idealized image of a school system which was embraced by the mass of the community. Goodman distrusts what schools have become, but he holds to a model which he believes existed in the past and to which—with some updating to be sure—we want to return.[26]

Indeed even historians who have been skeptical and cautious in their reverence for public schools have rarely held out long. Mary and Oscar Handlin, for example, have asked for more precision in our attempts to understand the school's relation to mobility, but they firmly conclude, nevertheless, that "the development of the American educational system in the long run positively accelerated social mobility." Aware as they are of the potency of forces outside the school—they note, for instance, that schools might reinforce a success already established in the real world—they hold to their belief that there has been a wide-

spread diffusion of property among all but the very underprivileged, and that "the school in one way or another is basic
to it."[27]

Although intense, impatient demands—themselves a symptom
of the past success of the school generating present expectations—are sometimes wrongly interpreted, the school is still
relied on to do its old work well. The schools are credited with
having done well during America's greatest recent crises, a
major depression and two world wars. They moved forward
under the pressure of reformers and public expectation, as they
always have, to meet the demands of a technological age by
recognizing its needs and trying to satisfy them, just as they
did for the explosion of the industrial age sixty years ago: new
curricula, new organization, new professional preparation, new
ventures in popularization and care for the least privileged. The
past is a promise for the future and, as in the past, we may look
to the future with considerable optimism.

Educational progress in America, largely a function of the
growth and popularization of the public schools, has come,
Henry Steele Commager observes, "largely from the requirements of the American democratic experiment, though it had
been leavened by pedagogical ideas of European philosophers.
. . . No other people," Commager says, "ever demanded so
much of its schools. . . . None other was ever so well served by
its schools and its educators. . . . To the schools went the
momentous responsibility of . . . inculating democracy, material-
ism, and equalitarianism."[28]

2

If the Rosy Picture
Were True. . .

I have recounted the familiar legend of public school magic through the ages because I need it for reference. I propose to draw some implications from it that I think will be less familiar. We have never really had much interest in asking what the social scene should look like if that rosy picture I have described, the official history of our schools, were an accurate one. I want to consider the American city, where the demand that the schools fulfill their historic mission is most immediate.

If the public school were as effective an agency for social regeneration as it is believed to be, three things would be clearly observable in the wider society. In the first place poverty, underemployment, large-scale and long-term immobility for lower class groups would be primarily—almost exclusively—black. Granting the costs of the slave experience and its cultural aftermath, many of those who believe in the great school legend argue that the mass of black Americans still have a long way to go before they can be considered "assimilated" in the manner of other ethnic groups' integration into the social order. They haven't made it yet, they say, but they will.[1]

Indeed, the increased number of years black children attend school is taken as an indication of their parents' determination to earn entry for their children into the opportunities of the wider society and also of that society's willingness to admit them. Naturally the same historic disadvantages, better under-

stood now than they were previously—in terms of such psychological and sociological factors as self-esteem and matriarchal
households—clearly make for less than successful performance
in schools. But the door is open, it is said, and, with time, the
most deserving of the underprivileged will move through it and
upward.

Secondly, if the rosy picture were true, Puerto Ricans and
blacks, "the new immigrants," should by now be moving wholesale across the cities they live in and out into the suburbs. The
intra-city mobility of upwardly mobile groups after 1920 and
even more rapidly since World War II is well known. Most observers date the "new immigrants" frontier at 1920 for blacks
and 1950 for Puerto Ricans. For at least the former group and
to an increasing extent for the latter, homes and home-owning
should be increasing apace in areas other than those in which
they were first settled. Parents in these communities should be
self-respecting citizens, working at dignified jobs, making it possible for third-generation progeny to enter into the more affluent
mainstream. After all, it must be, according to the analyses of
people like Silberman, Handlin, and Havighurst in the early
1960s, the middle-class blacks' and Puerto Ricans' turn to
follow the white exodus into the suburbs.[2]

Reflecting this movement, the overcrowding of inner city
schools would be alleviated, and what overcrowding persisted
would be more evenly distributed. Overcrowded, low-perfor-
mance, ethnically segregated schools would be increasingly
rare. The city itself would be relieved of many of the tensions
and pressures which now characterize it. An ever-expanding
vanguard from the ranks of the underprivileged would join
previously successful groups on an equal footing, segregated if
they so chose, but able to command full access to opportunity
and respect in the fulfilled sweetness of the American Dream.

And finally, there would be the school itself—the place
where we should all be able to find our future at work in micro-
cosm, the place where there should be no doubts about what
the norms of our society are. I don't mean that schools would
as yet actually be delivering all the social mobility they have

promised. But we should be able to look at our public schools and see a place where there is little controversy over whether all individuals deserved—and were getting—equal opportunities, at least to the extent that any social institution can provide them for different children, at different ages, at different times.

If the rosy picture were true, we would see urban children being reminded over and over that they inhabit a learning environment dedicated to their own respective ambitions in a wider social setting, becoming more and more democratic and less and less class-conscious in ways that predetermine the success or failure of children. Teachers would not act upon conscious or unconscious assumptions based on their own respective class or ethnic background to covertly or overtly, subtly or directly punish the children for theirs.[3] Punish, of course, means much more than the infliction of corporal pain. It refers to having children compete with each other against standardized measures in a game in which most of the children "playing" must lose. Such settings punish all because of the hostility they engender. In a society founded and developed on consensus instead of conflict, the school would be a place to generate harmony and encourage good feelings. Such schools would all be equally well equipped with facilities, materials, and services, and they would be supervised by teachers dedicated to completing the alleged historic mission of the American public schools: the constant redistribution, on increasingly more egalitarian grounds, of the human and material comforts of an open democracy.[4]

But what if the black community is not unique at all, but merely the most vociferous and longest abused of a long line of poor people? What if the narrative of other ethnic group progress through the schools is a legend and the institutions of economic and political advancement have never served more than a favored few, compelling large numbers to live in poverty and social discrimination? What if schools were successful only to the extent that maintaining economic and social deprivation on a massive scale is, as H. Rap Brown called violence, "as American as apple pie"? Then, clearly, the rosy picture collapses. The hallowed public schools' grounds for immunity from serious criticism would be at an end.

Once we recognize that the schools' success with minorities has been staggeringly exaggerated, can we continue to assume that they will be successful in advancing the interests of non-white minorities? For example, while American social institutions seem to have permitted talented individuals to rise up the ladder of power, it is quite fallacious to believe that those institutions have allowed the same room for movement to under-privileged groups seeking to share power with those groups comfortably lodged in positions of power.[5]

The poverty which is so much a part of the nation's "affluence"—from Appalachia to small, non-black ethnic communities in Coney Island, New York City—testifies to the fact that blacks do not suffer alone, even if they have suffered longer and continue to suffer more. Puerto Ricans, Poles, Italians, Haitians, Hong Kong Chinese, and white Anglo-Saxon Protestants—whether comparatively recent arrivals in the nation's cities or the not insignificant remnants of earlier migration to the cities—all perform a notch or two less well in school, are a degree or two less motivated, and will do increasingly less well in society given existing opportunities.[6] Among these poor one finds, too, a body of the more pathologically poor—the "underclass," as Gunnar Myrdal calls them—who often get lost in national bookkeeping which is now particularly and belatedly sensitive to black deprivation. They, like the ordinary poor, have not moved far from where their families settled, but more often the only mobility they know is that which has always, as Stephen Therstrom points out, been the lot of the abjectly poor —they move at the whim of the employer, of the welfare agency, of the police, of the credit collector; they are, and have historically been, lost to our various statistical assessments of social progress. The realities of their suffering have had virtually no influence on our good feelings about the inevitability of our social justice and progress. To neighboring communities, as well as to those who come to help, these people seem to deserve a level of scorn, abuse, and pity which cuts right across the heterogeneity of component ethnic parts. Teachers and administrators daily record, with varying degrees of frustration, the high correlation between abject school failure and the environment of despair.[7]

There are more than 45 million children in the public schools of America. They have many things in common. But what is most important is that both the children and their schools are, as Peter Schrag writes, "as different as suburbia and Appalachia, Nob Hill and Harlem, poverty and wealth, ambition and apathy."[8]

Some simple facts show the extent of the present troubles and the powerlessness of schools to ameliorate them in fifteen of our largest cities, New York, Chicago, Boston, Los Angeles, Philadelphia, Detroit, Baltimore, Houston, Cleveland, Washington, D.C., St. Louis, Milwaukee, San Francisco, Pittsburgh, Buffalo.[9]

—31 percent of the children who completed ninth grade in the big cities failed to receive their high school diplomas, against 24 percent nationally.

—In one city, the rate of unemployment for male school-leavers sixteen to twenty-one is 15 percent higher than the rate for high-school graduates.

—In the same city, 48 percent of the boys sixteen to twenty-one *with* high-school diplomas were unemployed; this pattern is found in other cities.

—While difficult to demonstrate statistically, it is clear that a great majority of school-leavers and a large percentage of high-school graduates are unemployable save for unskilled jobs which are rapidly disappearing.

Nationwide:

—More than 9 million children now enrolled in public schools will enter the labor market as functional—for job purposes—illiterates.

—One in four high schoolers drop out before graduation. Estimates vary, but between 25 and 50 percent of those who complete high school are menially employed or unemployed.

Since the census reports of 1960, which showed that more than half the population of the United States was living in metropolitan areas (61 percent of the population), these areas have continued to grow, and faster than the rest of the country. Metropolitan development has produced or intensified many social problems, most of which have had repercussions in education. The net effect has been to make the educational system reflect the distance of the society from its democratic goals.[10]

One major problem in the cities is intensified economic and racial segregation of the population. As the total population of a city center grows, the slum belt around the central business district becomes wider. This results from the growth in total population, but also from the concentration of lower-class people in areas of poorest housing, usually in the oldest parts of the city. Those who can easily afford to move away do so quickly. More gradually—but steadily—working-class people whose income permits it move out of the slum district and take up residence farther from the center of the city, while people in middle-class districts of the central city move out to middle-class suburbs. Thus the ever-growing total population divides itself into a lower-class concentration at the center, with successively higher socioeconomic groups at greater distances outward, and the upper middle class and the upper class inhabiting a ring around the cities—the suburbs.[11]

One major result in schools is the segregation of lower-class boys and girls into elementary and secondary schools where they are exposed only to other children of similar socioeconomic status. They are effectively denied the companionship, stimulation, and free interaction across classes which the public schools promise as a prelude to the open society.[12]

A study made by Patricia Sexton in Detroit, for example, illustrates the pattern. She obtained the average incomes of the families living in the various school districts and then grouped the 243 schools by income rank. The schools in a given income group tended to be located at about the same distance from the central business district. The schools that had children from highest-income families were farthest from the center of the city. The schools in the lower-income areas had poorer records of achievement, intelligence, and behavior, and a higher dropout rate. The schools in the higher-income areas had more pupils chosen for elementary and junior high school programs for gifted children and more students from senior high school who were going to college. Also, the schools closer to the center of the city had a higher proportion of families with mothers working and with mother receiving aid for dependent children.[13]

Nor are the problems diminishing. Hostility and competition between groups are intensified by frustrating battles over desegregation and decentralization. When the smoke clears, power lies where it always has.

The public schools have never really embraced the mass of the community, nor do they now. There is no point in reviewing here once again the frequency of academic failure for poor children, or the joylessness in middle-class schools, the over-riding fear of social disruption, or the unending sacrifice of the individual to the imperatives of economic growth. Still, we assess and plan, guided by a legend, believing all the while that we are making rational plans for the future.[14]

Our proper obligation is to make the schools *work*. To do that we have to acknowledge that we have no models, that we are asking the public school to adopt a new role. We must cease being satisfied with liberal rhetoric. The question is whether Americans are prepared to include poor people in the fulfillment of the national dreams.

Rarely, if ever, have the public schools become channels of substantial change either within their own structure or in the structure of society at large. But we have believed that they have, and we continue to believe it. Meanwhile, as universal public schooling adds more and more years to minimum required attendance, so higher education, too, has become an integral part of the joint popularization-standardization process. Ironically, "school" is also an American idiom for college. We have come to require school attendance, not require school success. As a result, more proletarians have had more years of schooling than ever before, but the phenomenal rates of failure they used to sustain at the elementary school level have shifted to the high school and college levels where the familiar 40 to

60 percent of all entrants fall behind, fail, and drop out.[15]

A most recent example of this process and typical of it is the question of opening up higher education to universal entrance. The G.I. Bill of Rights initiated the trend which has rapidly made it clear that higher academic achievement was to be increasingly essential for employment—either because it was truly preparatory or socially necessary because of shrinking manpower needs at some levels and increased selectivity at others. Whatever the reason, although most recent evidence seems to favor the latter, it has been clear for some time now that new layers would be "needed," and would be added to the top of the educational apparatus. As in the past, the addition of layers popularly represents reform—that is, the accommodation of society to new social-industrial pressure. No redistribution of opportunity or rewards is intended or received.

As Jencks and Riesman argue, "the whole labor force has been upgraded, but the relative position of various strata has evidently changed very little." A wide range of sources now confirms that relative mobility has not been great since the turn of the century. The rate has been much the same, in fact, as in other industrial societies, and the size of various social classes is essentially the same even though college attendance has expanded rapidly. The overall rise in higher education, especially in the last twenty-five years, is among middle-class youth. Opening up new levels between high school and college for those previously not college bound seems to mean widening the gap between social classes via education, at the top now rather than at the bottom.

The growth of enrollments in higher education coincides with the growth of an economy requiring less and less unskilled labor, and expands the capacity of the educational system to legitimize the wider system of class stratification. Perhaps, then, it is a mistake to view dropout rates in junior and senior colleges as a failure of the movement to democratize higher education, but rather as a successful adaptation of the educational apparatus to the continued performance of the task it was set up to perform.[16]

In New York City, for example, open enrollment has been hailed as continuing the democratic thread—though there were certainly enough rumblings and grumblings about the threat to standards. The educationally neediest groups today, the blacks and Puerto Ricans, it is said, are now being invited in. This epochal change is represented as a response—an overdue response—to blacks. At last public education can again be made "relevant to the world of here and now." And as usual the stated fears that the invited guest would be unable to cope with the new demands prepares the ground for blaming him when he does fail at this level of promotional sleight of hand. The reality hidden by the more popular story is that over 50 percent of those entering New York's City University under open enrollment are still white—most often first-generation college youngsters, the sons and daughters of Italian and Irish Americans who have only now begun to make slight progress up the socioeconomic ladder and for whom the expansion of white collar employment means "higher education" as a prerequisite for employment.[17]

We have been accustomed to taking Marcus Hansen's assessment that "the immigrant problem went away" as a truism without taking seriously the qualifying and dominant part of his statement. "The immigrant problem was not solved," Hansen wrote, "it simply went away." It went away as vocal vanguards from the successive immigrant waves left the ghettos, always leaving a large segment behind amid still later newcomers, new groups which come to be thought of as the only poor in the society. Immigration restriction (in fact a testament to the non-solution of the problem), our reduced awareness of the existence of poverty-stricken remnants of immigrants as social problems (aided by the dropping of "immigrant" as a variable in census collection), and the great migrations of blacks out of the South to Northern cities since 1917, all these have changed the appearance—the color, to be precise—of social problems from that of the equally unresolved problems, in almost identical form, which existed fifty years ago.[18]

But the legend has it that the public school system long ago provided equal opportunity for the parents and grandparents of these white "ethnics" now entering the city universities. It is a legend which deserves a closer look—and if it does not stand up to scrutiny, then it is a story we really ought to stop telling the children of the poor today. "Inspirational" lies are in fact a good deal less effective—not to speak of less fair—than they are cracked up to be.

3

The Historians
and the Legend

There is, then, a great incongruity between the history of public
schooling cherished by this nation and the reality we have inherited. Our first school historians erected the comforting story
and more recent historians have been largely responsible for
perpetuating it. Since the turn of the century, historians of education have moved from viewing the establishment of public
schools as the triumph of an enlightened working/lower middle
class over a selfish elite to a view which stressed a consensus in
all parts of American society on the principles governing the
essential place of schools in a democracy. But old and "new"
historians have generally been united in their conviction that the
national and ethnic group progress in which America takes
pride has its roots in the potency of the public school.

For the historians, as for most Americans, public education
is a religion, both because it serves a greater deity—the American Dream—and because public education itself is expected to
be omnipotent and omnipresent. Our historians' faith is a quasi-
egalitarian one which, through bittersweet half-truths and
supported by the unceasing propaganda produced by our
educational establishments, has been able to retain a working
relationship with a quite undemocratic social reality in both
public schools and American society as a whole. This faith,
like most religious systems, teaches that humanity—at least in
America—is quite properly organized into social tiers within
which men do the Almighty's bidding by fulfilling the functions

of their tier without too many questions. In addition, the underlying understanding regards a place on any one tier as no more
nor less than a chance to rise or fall into another one. The public school conventionalizes the faith as does any church, beckon-
ing some into the state of grace it reveals to them while pinch-
ing and poking all those traditionally squeezed out and left
upon their scraggy knees, as they are by all the other organized
religions of the world.

So we are brought to see the futility of attempts to resolve
our social problems using the schools—as they are—as our
major weapon. Our frame of reference is both abstract and all
too concrete, attitudes characteristic of those held by men of
strong religious conviction. Like any shrine, the public school
is looked to for workaday maintenance and for periodic miracles.

Despite increasingly sophisticated analyses, contemporary students of the growth and role of public education in America
have, with few exceptions, continued to write pretty much the
same story as that told by less sophisticated students equally
enamored of the cultural harmony and economic mobility they
perceived in their own eras. Despite the use of quantitative
techniques, infusions from psychology, psychoanalysis, and urban sociology, there has been no drastic revision of our view of
American educational history. As a result, despite an avalanche
of sharp criticism of schools, there have been few assessments
of the present that manage to discover the continuity that under-
lies the present school crisis.

The views of historical events we receive as conventional
wisdom are generally the effects, not the cause, of political be-
liefs. The historical search is invariably a search for actions in
accord with a predetermined rule rather than an effort to formu-
late a rule as an induction from behavior and the social reality
it expresses.

Looking for the underpinnings of society in abstract statements rather than historical circumstances helps to nourish a
huge American optismism. Whatever the messages coming up
from the depths of social and economic misery, promise dominates the nation's view of itself and raises to divine inspiration

the competitive achievement which charges it. The optimism—
the promise of opportunity and realizable economic mobility—
has taken on the aura of "fate," making personal philosophy
and national circumstances synonymous, just as any viable state
religion must do. The past is used in support of this national
optimism much as a salesman uses his fingers in the back of a
shoe to persuade the buyer that it fits.

As Heilbroner pointed out, social history and social conscience are companion outgrowths of the demise of church religion as the organizing base for men in society. School history
is a branch of social history which has been closer to church
history—written "in-house," with certain assumptions to be
taken absolutely for granted.[1]

These historians are our pious monks; their chronicles never
deviate from the optimism essential to the unchanging survival
of venerated institutions. They represent the hope Erik Erikson
observes: "there is always a hope [in the United States] that
in regard to any possible built-in evil . . . appropriate brakes
and corrections will be invented in the nick of time, without
any undue investment of strenuously new principles." More or
less of the same, over and over again.[2]

While I do not discount the theoretical and practical force of
those on the extreme right and extreme left who variously see
serious immediate social threats and espouse dramatic action, I
believe that the men of the middle-of-the-road, those persuaded
of progress and corrigibility—the disciples of rationality, compromise, and law—are the true interpreters of the word for industrial and post-industrial America—leaving the problems of
racism, poverty, and despair unalleviated, perhaps even exacer-
bated. Having inherited the right to enter the Holy of Holies
they protect it with their support of meritocratic privilege and
its efficient distribution by the public school system. They bow
to the values of individual growth and universal opportunity,
the rhetoric of the middle-class ideal, but they never judge
society by whether it has actually achieved those standards.

To interpret rhetorical good intentions as actual values, to regard them as actual priorities, is to read and judge a society quite

sympathetically. But it is not very good history or sociology.

Born in the age of Adam Smith's optimism and certitude, America proved the point by magnificent expansion. Growth was the watchword. There were problems, but it worked. The dark visions of Malthus and Ricardo, the fears of Mill and De Tocqueville, and the analyses of Marx were as thoroughly overshadowed by American growth as were the human costs being paid for the industrial, geographic, and demographic expansion. Even the Great Depression, which seriously challenged the nation's self-assurance for the first time, only dented it temporarily. The public school grew and changed in dutiful response to that expansion. It grew, as we shall see, out of the church's pedagogic role. It expanded with industry, urban growth, and ethnic diversity. Finally, as with the church before it, to question the public school's efficacy was to question both America's entire achievement to date and her future.

When they study education, most American scholars seem to make a working assumption about the society: that it is a competitive, fragmented society in which everybody, personally or in groups, has some power and nobody has or can have enough power to seriously unbalance the entire system.

Like classical economics, almost all contemporary social science (including history) is subordinated to the notion of scarcity. As Edward Shils saw it:

The acknowledgement of scarcity has been an essential element in the outlook of mankind over most of its history. Poverty and injustice, illness and brevity of life, the limitations on the possibility of gratifying desires and impulses have been regarded as inexpungeable elements of the situation of mankind and ethical patterns and theodicies have been constructed to justify or to ensure —and to integrate—this inevitable condition.[3]

Our schools are no exception. The assumption that there must always be losers, that achievement is proven only in competition is deeply ingrained in them.

The public school is the place in our society where the dualism between scarcity and optimism is theoretically resolved. But scarcity defines reality there as elsewhere in a society which

so deeply believes in the inevitability of scarcity. So reformers
—abolitionists, schoolmen, unionists, politicians, and academics
—in turn seek to accommodate their proposals for social change
(namely, the place of their ambitions in society) to what is
assumed to be the limits of adaptability of the prevailing social
order. As a result our secular ceremonies, like our purely religious ones, lack conviction, are devoid of seriousness. Throughout, the hidden agenda is to secure hierarchy and shore up
order based on it.

But the schools are locked in the dualism against which Dewey
inveighed. He saw a series of illegitimate oppositions as being
characteristic of education at all levels in this culture: knowing
vs. doing, freedom vs. authority, emotion vs. intellect, schooling vs. ignorance, patriotism vs. barbarism, nature vs. nurture,
and so on.[4]

The ultimate dualism for the school, its elitist reality amid its
egalitarian rhetoric, is resolved in the only way it can be, once
we assume the constancy of the economic order—by treating
erosions in the stated expectation of performance as isolated
cultural phenomena. The bootstrap theory is quickly reaffirmed
and the plight of those who cannot pull hard enough becomes
part of a closely related and equally unresolved dualism in the
field of social welfare: the sanctity of every life vs. inconve-
nience. Meanwhile, using the bell curve to sustain nineteenth-
century Wage Fund Theory (which argued in effect that the
industrial world was flat and, therefore, shares in it finite), the
public school selects winners and losers, more losers than winners, for it is on the effective exclusion of large numbers that
the security of an affluent community depends.[5]

For more than half a century, the tension between optimism
and the belief in the inevitability of scarcity has been resolved
through ingenious historical interpretations, and nowhere more
successfully than in the historiography of public education. The
ways in which this history is inaccurate I will discuss as I proceed, but its central role in creating the hallowed place of the
public school in American political, economic, and social thinking is indisputable.

The historiography to which I refer runs from E. P. Cubberley to Lawrence Cremin, from 1900 to 1965. The early
work was filiopietist, parochial, and narrowly institutional. It
was written in deference to the Whig historians' interpretation of
European and American events—an interpretation which saw
democracy expanding and which looked for its roots in the
past. Despite increased "scientific" rigor, with a more solid
basis in establishable fact, the latter has been no less a
vehicle for reaffirming the by-now longstanding conventional
wisdom that holds tightly to the efficacy of the American Dream
and the past glory of its sandman—the public school. The establishment and development of the school may not now be
viewed in quite the devotional terms embraced by Cubberley,
but educational historians, in attempting to be regarded as historians by historians, in no way questioned the school's greater
glory. Rather, the glorious success of the American public
school system has been the working assumption of post-Cub-
berley historical investigation.[6]

There has been comparatively little of the historiographical debate one finds in more popular fields such as Civil War history.
I do not mean to suggest that recent writing in the field is unimportant. Quite the reverse, it shows the degree to which
educational scholarship has been subservient to the social order,
and it shows what rich sources await the social or intellectual
historians who turn to public education as a way of examining
the structural essentials of that social order. But to the extent
that recent educational historical writing has been homogeneous
in both subject and approach, the history it has generated has
been wrong history, refusing to deal with the continuing social
reality—namely, that most of what we have come to believe in
respect to public schools and responsibility for American social
mobility consists of legends. Much of what passes for educational history in America exists to sustain that mythology with
interpretations generated by faith in the "Great School Legend."
The dominant early genre in the history of education in this
country was a version of house history comparable to the work

of early denominational historians. Like ministers writing sec-
tarian histories for their colleagues, educational historians often
wrote to unite and inspire their co-workers in the schools. From
the perspective of professional schoolmen, especially school administrators, they told a story of evolving institutions and increasing ideological consensus. Their narrative was linear, evolu-
tionary, and often used states and other administrative units as
a geographical focus, dividing the story chronologically according to the tenure of school officers. Commonly, school superintendents themselves wrote these histories. When others recorded
the past, they usually told the facts administrators chose to
divulge: statistics of attendance and enrollment, laws affecting
the schools, supervision, teachers' institutes, extension of the
curriculum, and related topics. In short, it was a promotional,
optimistic account of progress achieved; any remaining problems could be handled from within; the call, always, was for
greater efficiency along well-trodden paths. This was an insider's
view of the schools, seen from the top down, which dominated
the field right down through the Depression.[7]

Since then some changes have appeared in this establishment
view of American educational history. Historians have paid increasing attention to philosophical debates about the nature of
learning and the learner, about curriculum and classroom strat-
egies for democratic training. Yet few have asked, as did Merle
Curti, basic questions about the actual historical role and functioning of schooling, its structural relations, and the values
implicit in its "under-said and over-sung" results. Commonly the
history and philosophy of education were closely linked in de-
partments or colleges of education—indeed, often taught by the
same people. Thus it is not coincidental that much of the writing in the field centered on ideas about education rather than
on empirical investigation of actual practice. The apparent assumption was that commitment to a particular theory of education automatically resulted in behavior appropriate to it in the
classroom. Problems of personality, class, and hierarchial status,
and the question of the quality of commitment, went largely
unattended.[8]

I would like to deal briefly with two influential works: Bernard Bailyn's *Education in the Forming of American Society* (1960); and Lawrence Cremin's *The Genius of American Education* (1965). In different ways, these books made two central methodological points: (1) that education is far broader than schooling (indeed that "in a profound sense, it *is* the life of the young as they move toward maturity"); and (2) that historians should analyze the impact of education upon society, not simply "the character that education has acquired as a creation of society."[9] The Committee on the Role of Education in American History put the latter conviction in this manner: "the historian may . . . and we trust he will . . . approach education saying: here is a constellation of institutions . . . what difference have they made in the life of the society around them?" The Committee urged scholars to investigate the influence of education upon such matters as the assimilation of immigrants, economic development, equality of opportunity, the development of political values and institutions, and "the growth of a distinctive American culture over a vast continental area." As the assumption of national culture implies, however, the substance of that imperative became no more than a cue to address such significant questions with a view to understanding the casual role of the school in the successive achievement of those objectives.[10]

In *Education in the Forming of American Society* Bernard Bailyn proposed the writing of a history of education in the Colonial and Revolutionary periods which would take into account changes in the family, religious life, race relations, and economic development as crucial determinants in the transmission of culture from generation to generation. Rather than tracing in linear fashion the institutions of the present from seeds in the distant past, Bailyn urged historians to regain a sense of surprise, a sense of discontinuity. The transformation of education in America, he wrote,

becomes evident . . . when one assumes that the past was not incidentally but essentially different from the present; when one seeks as the points of greatest relevance those critical passages of history where elements of our familiar present, still parts of an unfamiliar past, begin to disentangle themselves, begin to emerge amid confusion and uncertainty. For these soft, ambiguous moments where the words we use and the institutions we know are notably present but are still enmeshed in older meanings and different purposes . . . these are the moments of true origination.[11]

Lawrence Cremin continued the historiographical dialogue. In his influential *The Genius of American Education,* as in his essay on Cubberley, he urged historians to examine educational agencies rather than the school, to bring revisionist interpretations in general history to bear on educational questions, to exploit insights from the social sciences, and to place American educational history in comparative perspective.[12]

These three works (three if we include the Cubberley essay) constituted important points of departure for educational historians because they placed the school in the wider social setting. But they continued, whatever their self-consciousness, to treat the public school as a proven asset in the unquestioned progress of democracy in American life. Theirs is a new, more sophisticated form of originology—they seek the origins of what they assume exists—and, not surprisingly, what they find in the past supports the truth of their assumptions. So they assume that our schools are democratic institutions and then proceed to discover how they got that way.

Another effect of the insistence that historians examine the influence of school on society is to reinforce the belief that the schools have, in fact, been a most important influence on the society. Again it avoids the primary question: did the school change the society at all or was its primary function to support and transmit the values and practices of the society intact? Not only does the call imply a preparedness on the part of scholars to find such an influence and thereby to sustain the interpretations of older scholarship, but it ignores what is supposed to be the major Bailyn-Cremin contribution: that other institutions in society besides the school educate.[13] In addition, historians should certainly be aware of the immense difficulty of any social science research designed to assess the precise "influence" of the schools within the complex of social, environmental, and perhaps biological factors which are the substance of change. Until we have explored such methods and developed such tools the question is moot, the assumption held the reverse of logic according to the data available.

Further, by establishing the multi-faceted aspect of man's education in society, these students of education confused the one institutional issue of primary importance to us. They failed to examine what education in schools really meant in terms of the social goals which must, covertly or overtly, lie behind any system of public financing. Essentially they bought the theory of philanthropic, egalitarian motives. They took the school into society but forgot that it was a separable institution with particular objectives and considerations that might inform our knowledge of that society. In essence they gave us the other side of the same coin. We have yet to question the currency.

Bailyn's *Education in the Forming of American Society* and Cremin's *The Genius of American Education* and *The Transformation of the School* established categorically the importance of studying education and schooling in the development of American society as a serious way of understanding that society. At least that has been the view of both social and educational historians since.[14] Both men have put the stamp of what appeared to be a new sophistication in the historiography of education on a decade of students. But, sadly, the result of their work is subject to the very same criticism Bailyn once leveled at the earlier historical establishment in the field.*

In *Some Historical Notes* Bailyn says of Ellwood Cubberley, as well as of Paul Monroe, Henry Suzzallo, and William H.

---

* Interestingly enough, education history followed some way behind more general social, diplomatic and political history. Despite following the standards of rigor and social scientific sophistication established by these branches, educational historians did not at the same time follow the various revisions of the American past initiated by them. Educational historiography remained wedded to the image of American egalitarianism and humanism, not at all sharing in the critical awakening, among scholars in other fields, to the idea that America's domestic and international present seriously challenged the traditional historiography which pictured America as the egalitarian champion of freedom and excellence both at home and abroad.

Kilpatrick: "They were men of great accomplishment, but they were moderate and measured in manner; controlled, methodical and rather humorless. They had glimpsed the promised land, and they pursued it with a passion. Embodiments of the Protestant ethic, they became fantastically successful academic entrepreneurs. By World War I they were the captains of a vast educational industry." Bailyn notes that their concern for professional recognition among historians took a back seat to the drive to give education—philosophy, sociology, history—the full status of an autonomous academic discipline.

"These writers," Bailyn goes on,

deeply involved in the contemporary problems of education and convinced that history contained a special message bearing on these problems, knew well what the outcome of history had been, and they sought in the past the early evidences of that conclusion. . . . What triumphed, what had to triumph, was the concept and practice of public education as they knew it: the victory, that is, of free, publicly supported and publicly controlled institutions organized into state systems and containing three distinct levels of instruction, elementary, secondary, and collegiate.[15]

That had been where Cremin was, as he later recognized, in 1951 when he wrote *The American Common School*. A little more than ten years later, both he and Bailyn had come to see the public school as only one of a number of institutions which educated children, and yet in their historical studies, they continued to make such large claims for the success of the schools that they never did justice to the impact of what the society taught both children and their teachers. Their differentiation between education and schooling is always imprecise at best.[16]

Both Bailyn and Cremin see the intellectual flaws in a history which is obviously little more than the story of the public school realizing itself over time, a history in which "colonial education becomes an anachronistic tale of origins that too often begins and ends with the Massachusetts 'Old Deluder Satan Act' of 1647." What they do not see, it seems, is that despite much tighter, better researched and documented reasoning, they too have produced the same history—only the institution whose divine origins are to be researched is the democratic American state, with the school as an important lever.

That is really no change at all. The "new" school histories of Cremin and Bailyn share the same serious intellectual flaws of the historiography they have replaced. Men and events are selected on the basis of their importance to the emerging, improving edifice, to the exclusion of men and events which underscore the origins and development of a restrictive, exclusionary, exploitative capitalist system and its institutions.

The historical method by which all events can be validated by locating their roots in previous periods is really no different from the Cubberley in-house history of education they both condemn. It is self-congratulation as history, looking to the past to show the inevitability of what is believed to be good and typical of the present. It is part and parcel of a superordinate democratic theory which depends on a congruent mood—"deeplying sentiment among adherents," as A. O. Lovejoy put it. Meanwhile, the view that every group will get a fair share, in good time, eliminates all possibility that the aggregate economic wants of a community might be satisfied.

None of these criticisms is meant to deny the validity—even importance—of defining and redefining even a wrongheaded but dominant point of view—if you happen to be a scholar and a subscriber to that point of view. The problem is that we have accepted it as a truly *revisionist* view of American education, which it never was in any substantive sense, and so again we have left ourselves with "no historical leverage on the problems of American education."

Bailyn looked to Colonial America for the roots of public education.[17] Recognizing that the school was a trivial subject for study when divorced from society, Bailyn examined its growth out of what he considered the uniquely open conditions of the American frontier and the culture shock by which he characterized the adaptation of immigrant people to American opportunities, dangers, and threats. In conceiving of education in broad terms, Bailyn attended to such out-of-school educators as family, church, and community. That emphasis generated some apparently provocative theses: that formal schooling accepted new cultural burdens because of drastic changes in the composition and character of the family in the new world; that with land abundant and labor at a premium, apprenticeship rapidly surrendered its traditional educative functions; that the initial effort to convert the Indians to Christianity introduced a new social role for the schools that persisted with the continued heterogeneity of the population; that public support for education resulted not from ideological principles but from a chronic lack of surplus private wealth; and finally, that by the end of the eighteenth century American education had undergone a radical transformation that can only be grasped as part of a larger transformation in the social, economic, and religious life of the colonies. Provocative yes, but quite untested against alternative assumptions and negative evidence. All that seemed relevant were those tales of progress which confirmed the desired conclusion that the schools were increasingly responsible for the expansion and perfection of an assumed American democracy.

It might be useful to question some of these assumptions. If the school did inherit the pedagogical functions of church and family, does that in fact necessarily mean that the family declined in importance? It might instead mean that the social pattern in which religion once rewarded privileged families with grace and power was replaced by a new secular institutional structure which administered formal education. But those families which suffered in one system suffered as well in the other. It seems quite probable that a more accurate history of the oppression, dulled hope, and economic immobility which have come to characterize our modern urban society can be developed out of the very sources Bailyn examines to congratulate our society for its accomplishments.

For example, the rise of formal schooling, the increasing dependence on schools for education, Bailyn argues, follows the breakup of the family as an intact, stable educational unit as a consequence of the culture shock of immigration to the New World.

He interprets the record of fear of family dissolution—of filial disobedience—contained in the laws established to punish its occurrence to mean that such dissolution was in fact already in progress. It may not have been so. Whether such laws were any more than one more expression of the anxieties of a pioneer population must remain an open question. It must be asked, at the very least, whether there is to be found, in the readiness of these pioneers to execute children for persistently disobeying parents, an important indication of the real principles involved in the beginnings of formal schooling.

In like manner, Bailyn interprets the work of the Colonists who zealously pursued the conversion of the Indians as another early step along the road to inclusion and opportunity for dispossessed minority groups through formal education.[18] Yet judging by the record, such efforts were not at all followed by Indian mobility or progress into the American democratic community. Christianizing was not unlike the Americanizing that followed it. It is worth noting that the Society for the Propagation of the Gospel working with Indians tried to do similar work with Negroes and latterly even Pennsylvania Germans "unfamiliar with English government." What mattered was conformity to a single standard and obedience to a single authority. Indeed, it is related to the simultaneous frenzied demand that all children always show recognizable respect for their elders; fear for the stability of society came above all else. The highest priority was given to protecting the status quo—its values, its opportunities and respect patterns, and its ethnocentrism—from serious challenge. Schools may have won tax support for very pragmatic financial reasons, but it is dangerous for both our understanding of the past and the present to miss the highly ideological component which from the beginning accompanied rhetorical Christian concern for proselytizing on behalf of God and brotherly love. In the New World this political-religious equation was expressed by the belief in futurity woven inextricably with the other side of Puritan thought—belief in the scarcity and material manifestations of the state of grace.

Although Bailyn is persuasive, he misleads us about the nature of American society as well as of American education. He is convinced, on scanty evidence, of the decline of the family. After all, the rise of the nuclear family is not decline in the influence of the family, and there is evidence to suggest that for early settlers geographic mobility was actually small from generation to generation. On the basis of his conviction that the family declined, Bailyn asserts that "schools and formal schooling had acquired a new importance. They had assumed cultural burdens they had not borne before." And assumed them successfully, he believes.[19] They did so only if one measures their success from the vantage point of Puritanism's exclusionary distribution of grace rather than by the democratic promise that salvation would be available to all. Indeed, the school's enshrined limitation defined it for an expanding industrial nation, while adopting the same shallow democratic promise and making it the official school language we are peculiarly bemused by today.

It is clear to Bailyn, but for no reason that is made apparent in his book, that the Colonial educational inheritance he has examined "released rather than impeded the restless energies and ambitions of groups and individuals." The processes of education, formalized and sanctioned "in a framework of enlightened political thought," tended "to isolate the individual, to propel him away from the simple acceptance of a predetermined social role, and to nourish his distrust of authority." Finally Bailyn's own effort to treat non-school education gets blurred as he progresses. When he discusses the state-supported school, he seems to abandon all those extra-school factors he discovered. In his concluding sections he completely confuses the difference. He almost exclusively talks about the public school when he speaks about education turning back and acting on society and the transformation of education that took place after the Revolution.[20]

As Bailyn's confusion on this issue indicates, the school was increasingly gaining a monopoly on selection for opportunity. The public school did not simply replace the family but forced the less than middle-class family (defined by status or aspiration level) into a state of dependence upon it, on its accrediting standards, on its dominant class norms. The family was not replaced but rewarded and punished differently, and by a "new" agency. Bailyn forgot the social scientific knowledge which goes along with the social scientific awareness of the importance of family background—its respective potency and weakness in the dominant culture for rich and poor, and the inability of social institutions to do more than reinforce those opportunities where they are.[21]

Overly concerned with the task of discrediting Cubberley's simplistic notion, which in a very literal-minded way traces the public school back to a model existing in the seventeenth-century Colonies, Bailyn overlooks his own dependence upon ideas about continuity—in his case the continuity of democratic purpose rather than institutional forms. He recognizes the birth of a new structure: "Public education as it was in the late nineteenth century, and is now, had not grown from known seventeenth century seeds: it was a new and unexpected genus whose ultimate character could not have been predicted . . ."[22] But he assumes that both seventeenth- and nineteenth-century American educational structures are steps along the long but open road to equality and social justice. Instead, formal education has been—from its beginnings—an agent for defining and limiting American community.

Most certainly, as Bailyn points out, the school was "part of a complex story, involving changes in the role of the state as well as in the general institutional character of society. It is elaborately woven into the fabric of early modern history." But so are the realities of Colonial America. The growth of the Colonies and the emergence of the United States pinpoints what is truly unique about America—the growth of a nation state on a sustaining Puritan ideology and on industrial, capitalist economies which built from scratch what Britain had achieved over a long and turbulent history.

In his preface to *The Genius of American Education,* Cremin says firmly that "The thesis of this volume is that the genius of American education—its animating spirit, its most distinctive quality—lies in the commitment to popularization."[23] Because underprivileged groups looked to schools for help (Cremin cites Negroes in this essay) and the establishment promised help through the schools (he cites businessmen and educational professionals), he is prepared somehow to argue that such help was received and the public schools gave it. No clear distinction is maintained, throughout Cremin's major work, between "philosophy of education"—that is, intentions and rhetoric—and the reality of what schooling in the real world is like.

Cremin too recognizes that the school divorced from society is a trivial object for study. "The classic treatise (on education), of course," Cremin says, "is Plato's *Republic,* which remains to this day the most penetrating analysis of education and politics ever undertaken. Recall Plato's argument: In order to talk about the good society, we have to talk about the kind of education that will bring that society into existence and sustain it."[24] Apparently it is quite unimportant that Plato's educational vision was based on a concept of the good society which excluded large numbers of the citizenry and indeed depended on that exclusion for its existence; education, broadly conceived, was, for Plato, the major vehicle for sustaining such a social order. Those Platonic notions that have value for our understanding of the role of education in society and of the society itself through its system and principles of education are not considered. "Here there is," Cremin notes, "no vision of the good life that does not imply a set of educational policies; and conversely, every educational policy has implicit in it a vision of the good life."[25] But he does not go beyond the rhetoric. The "good" life for whom and defined by whom? What is the American vision as we can derive it from educational policies? Why does Cremin allow statements of educational ideals to stand for educational policies and practices? They are, after all, quite different.

"Recall, too," Cremin continues, "that when Plato gets around to talking about education, he gives relatively little attention to schools. As far as Plato is concerned, it is the community that educates, by which he means all the influences that mold the mind and character of the young: music, architecture, drama, painting, poetry, laws, and athletics."[26]

Is there then no "educational effect" exerted by what is present in the society which is *not* high culture? What of poverty, slavery, crime, cruelty, and indifference as educational influences on the young? Cremin never really explores this. He remains content to define community influences as those derived from high culture: art, poetry, and law. It is a fatal blind spot in a historian of education. If only "good influences" are considered part of a child's education and "bad influences" are not considered a part of his education, then there can be no debate about the efficacy of the schools as long as we find enough of what we are looking for, "good influences."

"In our own country," Cremin elaborates, "it was Thomas Jefferson who first articulated the inextricable tie between education and the politics of a free society. . . . We are all familiar with the proposal he made to the Virginia legislature for a state education system that would offer three years of public schooling to every free white child of the Commonwealth and then send the brightest youngsters on to grammar school and college free of charge."[27]

It never seems to strike Cremin as in any way odd that Jefferson, the apostle of the faith that education is the route to achievement and equality, was content in his most visionary plans for the Republic to exclude immense numbers of his countrymen, white as well as black, from real educational opportunity and achievement. There were, of course, other routes to full participation in late eighteenth- and early nineteenth-century society besides formal education, but the point is that for both Cremin and Jefferson education is considered *the route* to fulfillment, both personal and social, and yet they found it quite "natural" and acceptable to assume that equal educational opportunity for the "gifted few" was sufficient justification for the continued claim that education in America was truly democratic.

Nor does Cremin fully deal with the inconvenient fact that Jefferson's plan, with all its shortcomings, was turned down by the Virginia legislature. This is a sadly representative illustration of our historians' tendency to substitute a history of ideas, the record of proposals and philosophies, for actual school history.

"Jefferson's plan was turned down," Cremin admits, "but one can trace an unbroken line of influence from Jefferson to Horace Mann to John Dewey—and to trace it to Dewey is, like it or not, to trace it to ourselves."[28] I would add that this is true only if we are tracing the history of generous proposals which were turned down.

In *The Transformation of the School* Cremin examined the progressive movement in education at a time when the urban public schools were faced with rapid industrialization, urbanism, and large-scale immigration.[29] In this work, he argues that the progressive movement—directed, as Cremin sees it, to improved standards, more equal opportunities, a wide popular base—dominated the development of the school as we know it now; it was a part of a wider-based humanitarian, democratic spirit in the nation. *The Transformation of the School* is a much more rigorous work than *The Genius of American Education*. The rigor of its analysis and documentation, within the limits of intellectual history and the assumptions I take issue with, are impeccable. However, it maintains—as does Cremin's most recent and equally scholarly work, *American Education: The Colonial Experience*—the same assumption of America's essential egalitarianism and reflects continued optimism in that direction based on past successes attributable to the public school.[30]

In order, for example, to make Americans of immigrants and provide mobility and material well-being for untold millions of the dispossessed, ravished, and poor, the schools were themselves "inexorably modified," Cremin believes, into institutional symbols of the integrity of America's democratic commitment. Cremin sees in the development around the school of the new welfare services we have come to associate with urban poverty and in the growing concern of schools with "immigrant education" a desire on the part of people closely associated with schools to improve the quality of life for urban masses. Despite his awareness of the segregative nature of public schools for the upper classes and the not-so-common nature of schools for blacks in the South, the legend of "the single-class slum school that brought together immigrant children" is not to be questioned. Massive poverty and monumental immobility are nowhere at issue; he has no doubt that the school, at least in this century, has been vigorously involved in "the struggle for a better life."[31]

Ready to acknowledge that school reformers were moderates, not radicals, Cremin points out that "for all their sense of outrage, moderates take time," although he somehow excuses the fact that the time seems never to come. The ultimate dichotomy between conservative caution and social improvement on a large scale is always left quite unexamined. The possibility that there is a relationship between the conservatism he perceives and the lack of wide-scale, structural, social progress in America goes unconsidered, so that the assumption of the promise of American democracy retains an illogical but powerful hold. "The real radicals of the nineties," Cremin writes,

men like Eugene Victor Debs and Daniel De Leon—had little patience for reform through education. . . . But for the much larger group impelled by conscience yet restrained by conservatism, education provided a field par excellence for reform activities untainted by radicalism. . . . A half century earlier Horace Mann, certainly no radical, had refashioned the school as an engine to create a new republican America. It should hardly be surprising that a generation which followed him would again view education as an instrument to realize America's promise.

It never seems to strike him that perhaps the "promise," like the "new republican America" of the mid-nineteenth century, was drastically and quite unresolvably different from the radical vision of social regeneration. He remains convinced that early twentieth-century efforts to apply scientific method and social philanthropy to classrooms which were to include more and more children from hitherto excluded parts of the population were fundamentally "part of a vast humanitarian effort to apply the promise of American life—the ideal of government by, of, and for the people—to the puzzling new urban-industrial civilization. . . ."[32]

Progressives were "moderates" then, Cremin concedes, horrified by urban conditions and devoted to ameliorating them, but reform took time. Nevertheless, having studied the record of their expressions of horror and the various steps they took to change things, he is convinced that the reform route was a sure one.[33] Meanwhile, the high degree of school failure and "unspeakable" school conditions which progressives found in major cities at the turn of the twentieth century have continued to be the experience of the majority of urban school children. Somehow the fact that this majority has been increasingly non-white since World War II has diverted attention from the constancy of urban school and urban slum environments.

The school's positive work with immigrants is central to both Cremin's and Bailyn's faith in the public school, yet neither deals with the question in any detail. Since Cremin works largely in the world of ideas and intentions, we are never invited to consider evidence of either the immigrant's or the school's actual experience with each other and American society. Bailyn deals with the Colonial period but he closes his work with conclusions about the nineteenth century which are not history but his contemporary observations and biases. And yet, it is this perception of immigrant advancement in America, in part perhaps informed by each man's immigrant heritage and academic mobility route, which seems to underlie the fundamental optimism of their respective historical analyses. Cremin sees the school changed drastically, "transformed," in the face of large-scale immigration in order to fulfill its promise to the millions who brought to it a new assortment of needs. It never becomes necessary in his analysis to consider whether or not the changes in schools he identifies had any real impact on the mobility of those who attended. The impact is assumed, and it is a positive one. Bailyn assumes the same nineteenth-century transformation. The facts of urbanization and immigration make the importance of the late nineteenth century obvious. But Bailyn considers that it was the transformation of education in this period,

somehow separable as a factor from the pressures which provoked the changes in it, that "significantly shaped the development of American society." Certainly schools were basic to the
social service revolution getting under way amid rapid urbanization, and to be sure schooling has become an increasingly important prerequisite to job opportunity. But to assume on the
basis of unstated evidence that the modern public school has
been "an agency of rapid social change, a powerful internal
accelerator" which freed men from local group tradition and set
them on their way up the socioeconomic ladder is to read the
impact of the schools on immigrants at the turn of the twentieth
century in a narrowly nostalgic and convenient manner—a manner which leaves the current breakdown of schools with respect
to poor people, and the black poor in particular, inexplicable in
any terms other than to blame the poor themselves for their lot.

Both Cremin and Bailyn, it seems, are moved to their faith
in schools by what they consider, and perhaps what they have
experienced, to be the essentially egalitarian and ameliorative
nature of America's history and contemporary social and educational commitments. Ivan Illich's interpretation of the reasons
for the early trust and growing faith of Americans in formal
school education which Cremin and Bailyn identify so respect-
fully is much more logical in the face of existing social and
school inequalities. Illich makes much different assumptions
about the present and so it is not surprising that he draws rather
different conclusions about the past. According to Illich, for a
movement away from religious-based social organization to
nation-state secular organization to have been a step toward a
more humane and democratic society there should have been
no New England schools, Jefferson Plans, or Northwest Ordi-
dance, no established school systems at all. Schools are essentially a negative, regressive element in the early republic, Illich
says. Established schools appear early in American history, as
the data from our most recent educational historians show, out
of fear of change and desire for social stasis, not humaneness
or egalitarianism at all. "The first article of a bill of rights for a
modern, humanist society," Illich says, "would correspond to

the First Amendment to the U.S. Constitution: THE STATE SHALL MAKE NO LAW WITH RESPECT TO THE ES-TABLISHMENT OF EDUCATION. There shall be no ritual obligatory for all."[34]

For Illich, as for Bailyn and Cremin, the public school is a symbol of America. All three agree that public education was at the center of the movement from a religious to secular based society in the formative days of the republic. Illich, of course, does leave himself open to charges of historicism when he seems to be saying that public schools should not have been established in the first place since they are so bad now. But the important thing Illich has to tell us—and what distinguishes him from Bailyn and Cremin—is his conviction that public schools are what they are today as a result of their traditional, consistent ideological limitations. Here he is right and reads our history well. And yet, Illich's call for "deschooling," the eradication of compulsory public schooling, is finally as superficial in its lofty obfuscation of the motives and interests which make and sustain social institutions as the more traditional school analysts' devotion to the Great School Legend.

The public school was an extension and reinvigoration of the middle-class family ideal Henry Steele Commager described in 1949 in *The American Mind*—devoted to hard work, perseverance, respect for women.[35] The public school was not primarily designed to free the lower-class family from its low self-esteem and advance its members in society. It was an apparatus designed to control most of them, to safeguard society. Indeed, the relation of the poor family to the early public school is reminiscent and a precursor of the more dysfunctional relationship of the poor inner-city public school to the promise of public education. The view of the school as an agent of social regeneration is as unjustified as Commager's belief that the "American Boy" was indifferent to sex until adolescense and romantic about it afterwards. He may have been, but that is not the whole, unconflicted story as psychologists have begun to show us.

The history of the schools Bailyn and Cremin have written is

in essence the story of good boys who succeeded and of bad
ones who failed. It is never seen as possibly a history of those
who accepted the system and did what was expected of them
and of those who refused to do so or were excluded from making the choice—both were dropped by the wayside. Neither
Cremin nor Bailyn tells us of the men who put fear in the heart
of the elites, those in the lower middle and bottom ranges of
society who man machines, fight in revolutions, and then, for
the most part, man machines again. Nor about men of more
advanced ideas, men whose democratic thought rejected checks
and balances which were seen as a means of shackling the popular—and the individual—will. In place of the line that established educational historiography draws from Jefferson to Mann
to Dewey, why not a line which runs from Tom Paine to
Thoreau to Paul Goodman, committed to the release of individuality above all things, of openness for the human spirit?

What we have is an intellectual subsidy to maintain and re-
furbish the middle-class business ethic. Daniel Bell's definition
of democratic politics—"bargaining between legitimate groups
and the search for consensus"—is directly applicable to democratic schooling in America. The legendary public school, the
route to socioeconomic, even cultural, consensus, has validity
only so long as the analysis is limited by the exclusiveness implicit in the notion of "legitimate" demands.

The school stands today as a secular base in "democratic
mass society" for the Puritan ethic which supports industry,
urbanism, technology. Like the church before it, it stands for
the practical interpretation of ideals favoring stability and efficiency. The gap between rhetoric and reality in both the divine
and the secular allowed an elite to dominate in each society.
That is as true today as it was in the periods so optimistically
and selectively studied by the key figures of American educational historiography. In order to examine what really happened
in United States school history, we must uncover the intentions
implicit in the constancy of the results, rather than continue to
trust the rhetoric of intentions, and then ascribe the failure of all
those good intentions to mere "accident" or to "difficulty."

PART

II

What
Really
Happened

4

School Reform:
Liberal Rhetoric for
Conservative Goals

America is a very conservative society which likes to claim that it is devoted to equality and social change. It has a public school system designed to preserve that contradiction—by institutionalizing the rhetoric of change to preserve social stasis. The educational theory of its schools is a theory of a society based on scarcity; limited success and much failure is the educational equivalent of definitions of full employment that accept high levels of unemployment as tolerable. In fact the public school stands as an instrument of the conservative strategy for defusing movements for social change which seriously challenge the established order and it typifies social reform in the United States. This is the liberal tradition as Louis Hartz recognized it. Its conservative thrust has been, as Hartz so brilliantly analyzed, its most important effect on the democratic protestations of the American Dream.[1]

Even those who have disagreed with Hartz have to concede that American liberalism is a mass movement embracing different elements from different classes. It has a left, right, and center. "Liberalism is the mass Left wing of American society . . . ," Michael Harrington argues. And yet, if modern liberalism, and its immediate precursor American progressivism, defines that part of the nation sympathetically disposed to the democratic, egalitarian solution of social problems, then clearly the resolution of these problems requires solutions which go beyond liberalism.

Hartz's insight has led to a number of studies which have shown how, over and over again in American history, the ethical projection of ideals has been divorced from critical action in the real world.[2] Yet educational historians have not employed this insight to better understand the situation which they themselves, from an optimistic perspective, argue is basic to American social goals and social practice.

The essential conservatism of American reform movements may be illustrated by examining the common school movement and the abolition movement, both of which have been regarded as prime examples of the expansion of American democracy. In fact, the mid-nineteenth-century movements for the establishment of the "common" school and against slavery are marvelously clear examples of the essential conservatism of liberal reform. Both the school-reform and abolition movements were part and parcel of the contradictions I have talked about; they never fully intended the kinds of changes we have come to associate with the language they used.

I wish to include the abolition movement in this analysis of school history because many of its principal figures were prominent in school reform and both movements were responses to the pressures of industrialism and urbanism through which modern American culture was being redefined. In addition, the abolitionists are generally thought of today as uncompromising reformers, great fighters for truth, justice, and social equality; in fact, their legend is as pervasive, and as misleading, as that which surrounds the school reforms of the nineteenth century. Much is to be learned about liberal reform during the transition to industrialism by considering the two movements together. Slavery is inimical to the development of a consumer labor force for industrial society. To consume, goods workers must have wages. And school changes were clearly related to the appearance of large numbers of unskilled immigrant laborers who threatened the social order with their alien ways. These people had to be first contained and then socialized. In addition, the new industrial society needed a new kind of "white collar" worker with clerical skills.

I have said that American culture was subjected to redefinition by abolitionist and school reform responses to industrialization. I use the word "redefinition" carefully, because intrinsic to my argument is the belief that the treatment of successive American "revolutions" and "transformations" which run through traditional historiographical work profoundly distorts the available information. "Redefinition" implies change in order to retain dominant social relationships with no sense at all of the radical revisions "revolution" and "transformation" suggest.

"Transformation" is a consensus term. Conflict terms, the use of epochal descriptions such as "revolution," are relatively new and are generally used now only in relation to blacks. "Redefinition"—perhaps "reformulation" as Erik Erikson uses it— is a better word, meaning change directed toward maintaining the status quo and therefore occurring within the same encompassing theoretical and structural design. The capacity for such change, of course, has great survival value, but it does not by any means suggest the metamorphosis or relocation of direction that the word "transformation" implies. In this context social conflict is a source of adjustment to grievances without structural change.[3]

The problem with "transformation" is that it diverts us from the definition of progress implicit in the intra-institutional improvement on which school history is built. The changes which create an absolute increase in the number of those included in America's prosperous community are reverently praised; the relative consistency of the relative place of groups in the social hierarchy is not taken into account. For example, the inclusion of large numbers of students in public schools or colleges since 1900 has not meant a change in the lot or the relative size of the poor in relation to other ranks in the social order. Such a definition of progress in effect assumes the perpetual immobility of a relatively static lower-class base. Only a change characterized by a definition of progress which does not permit such relative immobility could be appropriately called "transformation."

Over and over again school reformers who have proposed improvements in public schools have talked of the desirability of warmer, more humane, less restrictive environments. But just as often, such reformers have ended up helping to produce formal educational hierarchies that constrict individual development and restrict political or religious controversy within the schools. The rhetorical concern of school reformers for the underdog invariably becomes compromised by the interests of those higher on the social ladder.

Maxine Greene, in a fascinating study which compares school reformers' visions of society with those of literary men from the beginning of the campaign for public schools to the end of the nineteenth century, has shown that the beginning of the movement to "establish" schools was characterized by such compromises.[4]

In 1839, after hearing Horace Mann deliver one of his talks, Ralph Waldo Emerson wrote in his Journal: "We are shut in schools . . . for ten or fifteen years, and come out at last with a bellyful of words and do not know a thing." To know, for Emerson, meant to feel his poetic imagination soar. It meant to open his soul to the "Oversoul," to see by the "Divine light of reason" with which every human being was endowed. The Common School, teaching conventional or "common" habits of thought and perception, seemed to him a barrier against authenticity. The school reformers, he believed, would make impossible the "self-reliance" which alone permitted God to enter through the "private door." If, as was likely, the school inculcated vulgar and self-serving habits, or the values associated with Trade, it would merely serve to perpetuate an inadequate society, an Establishment that was basically inhumane.

Horace Mann, the lawyer who became Massachusetts' first Secretary of the Board of Education, was a supporter of abolitionism, temperance, railroad building, hospitals for the insane, and tax-supported schools. Massachusetts and Mann were at the forefront of a by now widely based attempt to make local schools a matter of state concern. Mann told his audiences that "lower classes," coming together in cities now, were dependent on the conscience and funds of middle-class parents, that public schools would insure the maintenance of public order and the protection of property. We have been led to believe that he emphasized these concerns alongside and over his concern for the virtues of equality and mass enlightenment because he knew that the movement could not succeed without the support of the mass middle-class which was not primarily humantarian but self-interested. The fact is, his realism notwithstanding, the schools were to be the servants of those who built them. The same "realism" incidentally persuaded him to divorce his anti-slavery views from public school campaigning for fear of alien-ating the necessary middle-class taxpayers; "realism" and white supremacy won, in school and out.[5]

Mann talked, as did reformers in Connecticut, New Jersey, New York, and Pennsylvania, of the value of shared daily child-hood experiences to bridge the gulf between the classes; the schools would be "the great equalizer of the conditions of men," but it was the schools as "the balance wheel of the social ma-chinery" which triumphed—the balance being the imposition of controls for social stability in favor of the moneyed and powerful, and not the substance behind egalitarian rhetoric. In-deed, Mann wrote, "a government by mere force, however arbi-trary and cruel, has been held preferable to no government."[6]

Humanitarian reactions to nineteenth-century poverty never rose above a profound disrespect for the poor. This attitude has pervaded such activities in America since then, both public and private. What was new about poverty then was its relative den-sity in the cities and the dangers which were envisioned as a result. New York's Society for the Prevention of Pauperism an-nounced in 1821 that "the paupers of the city are, for the most part . . . depraved and vicious, and require support because they are so." Such declarations were repeated throughout the century in Philadelphia, Boston, Baltimore, and other growing urban centers. It was unanimously held that the chief causes of pov-erty were moral, and agencies designed to deal with poverty had also to impose strict moral standards on the poor they dealt with. The most vigorous expectations were always related to safeguarding the status quo and the rights to privilege and property. As David Rothman, the author of an important recent study of social outcasts in mid-nineteenth-century America, asserts, "Vice, crime, and poverty were stops on the same line, and men shuttled regularly among them." Somehow it was the community's obligation to protect the child, as New York Jacksonian philanthropists put it, "from the combined pressure of ignorance and evil example in the home where he was nursed and cradled . . . and the influences to which the street subjects him."[7]

As a result, it is generally assumed that the common schools grew up in their modern form around the middle of the nineteenth century in order to continue the nation's already established egalitarian direction. The school would provide more and more people with access to social and economic mobility while protecting society. In fact the latter imperative was always dominant. The school's continuity with the past was to be found in the fact that it reflected and reinforced what had been from the beginning the restrictive class nature of society. It supported class distinctions and was expected to socialize children for their places in the world. It was not surprising therefore that when the Irish famines of 1847 sent thousands of Irishmen to American cities, the first duty of the schools was to protect society from the "moral cesspool" they created in the cities by simply containing the newcomers, keeping them under observation, and subjecting them to the habits and values of their betters.[8]

The nation's growing demand for urban labor required a mechanism which would "balance" this need against the pressure of Anglo-Puritan social restrictions—its vicious hatred of Catholics and its immense racial and cultural bigotry. Indeed, if there had not been such a need for manpower—and an ever growing need at that—it is likely that the prospect of widespread pauperism (especially burdensome in times of slump) and the fear of papal power would have discouraged the mass entry of Irishmen into Northeastern cities whatever the faith that the schools would salvage and Americanize them. But they did come, and the schools, as Jefferson had hoped, proceeded to separate the best minds "from the rubbish." But selecting out the rubbish was always an important part of the job. Given also that there was apparently always more rubbish than talent, it seems important to recognize that at least part of the myth of the school's success grows out of a very recent reinterpretation of school success. Apparently the school system's success, until comparatively recently, meant recording the failure of large numbers of those entering, attending, or being truant from the public schools.

Henry Barnard, Connecticut's leading schoolman and later Federal Commissioner of Education, and Mann, after his resignation from educational office in Massachusetts after 1848, aptly characterize the place of schools in the nation. Mann accepted election to Congress and struck out vociferously against both slavery—on institutional, economic, and moral grounds—and the ruthlessness of surging capitalism. He got outside the schools because they were finally too slow a route to the social regeneration of which he dreamed. Barnard, a career educational professional, was much less explicit in condemning the abuses of the system; he was much more comfortable with the piecemeal change which went along with the security of economic status of the privileged classes. While Southerners learned increasingly to hate Mann because of his legislative stand against slavery, two Southern cities, New Orleans and Charleston, offered Barnard the superintendency of their schools. The industrious and frugal citizens Barnard's schools were expected to produce were needed on the Southern side of the Mason-Dixon Line too.[9] Indeed, the professional schoolman's reform stance, as typified by both men as school leaders, was much the same as that among abolitionists of the period in general, where it seems that antislavery for the most part had little to do with combating racism.

It is important to understand that what finally made antislavery a respectable movement was the inclusion of anti-Negro factions—racism triumphed over slavery. After all, as abolitionists realized in 1852, a society pervaded by an all but universal belief in white supremacy could not possibly permit equal rights for blacks. Slavery had to be separated as an issue from egalitarianism, even if defined within it, if the order which no longer had need of it was to be satisfactorily served.[10]

Indeed, it was the industrializing Northern states and the newly won frontier states which were the solid opponents of slavery. Southern states, much farther back on the road to competitive farming and urban industry, depended on an antiquated social selection system, not the principle of selection by race and class.[11]

Several recent studies seem to confirm De Tocqueville's observation in 1835 that "the prejudice of race appears to be stronger in the states which abolished slavery than in those where it still exists; and nowhere is it so important as in those states where servitude never has been known." In the frontier regions which became free states or territories by 1860, Eugene Berwanger shows quite clearly that anti-Negro activity intensified in the 1840s and 1850s coincident with the height of agitation over the issues of the expansion of slavery into the new territories. His evidence suggests that the migration of Negroes was a much greater concern than the expansion of slavery in the Middle West. From early in the nineteenth century, Illinois, Indiana, Oregon, and Kansas established a series of codes designed to discourage black residents. The fear of producing an eventual free Negro population escalated restrictive statutes from the threat of fines and lashes to the exclusion of the blacks from suffrage. Michigan, Iowa, and Wisconsin followed suit. Even in states where there were few blacks, identical statute restrictions were effected, as in Utah, Colorado, New Mexico, and Nebraska. Nebraska had only sixty-seven blacks in 1860, but black exclusion was still a rabid issue.[12]

In the Northeast, where vast majorities opposed abolitionism with hostility or indifference, fear of black competition, black political naïveté, and just plain refusal to countenance black equality with whites persuaded even men who opposed slavery to oppose abolition too. Politicians and clergymen were afraid of disrupting their established institutions despite a powerful rhetorical commitment to the regeneration of society on democratic egalitarian lines. School reformers like Mann had to keep their abolition position quite separate from school issues. The upshot of this was, of course, that the democratic school ethos and the faith in schools which followed it by definition excluded blacks. Indeed, until desegregation and attempts at equal educational opportunity legislation in the mid-twentieth century challenged this division, separate schools and separate classes were the tradition. Despite the overlap of personnel involved in the two movements, common school and abolition reformers refused—or were unable—to view and act upon the problems of education and slavery jointly. This was the first step in a series which led to the still persistent separation of the two problems despite their intrinsic relation in the existing social system. This separation contributes to the current inability of public education to deal with the large number of blacks in the schools, much less understand the black problem except in terms of the unfortunate heritage of slavery. Ironically, if school analysts included schools themselves in that heritage, their assessment of the black plight would be a more accurate one.[13]

Abolitionists claimed to be "realists" and believed that moderate support could be won in a society whose government and institutions were fundamentally sound. The fact that slavery was one of the institutions which was taken for granted by that society and recognized in its Constitution did not deter them. Unlike William Lloyd Garrison, the early but badly underestimated leader of the movement, abolitionists were, for the most part, directed by a belief in the essential workability and democracy of the existing social system. Anomalies would be smoothed away with time; survival and freedom from violent social disruption were superordinate goals for the reformer. Garrison, on the other hand, denounced the Constitution as a "covenant with death, and an agreement with hell" which "should be immediately annulled." Survival was not a primary consideration in radical logic.[14]

Garrison engaged in a theoretical rethinking of society, and for him abolitionism was a truly radical movement aimed at restructuring society through the elimination of exclusionary social bulwarks which in the North as well as the South generated racism. Unlike his opponents in the abolition movement (and his severest critics among historians) Garrison held to a radical, not a reform, line.

As Vann Woodward reminds us, despite the North's gradual abolition of slavery within its various borders by 1804, it "entered the Civil War as part of a slave republic to defend a Constitution that guaranteed slave property, led by a party with a platform pledged to protect slavery where it existed, and headed by a President who declared in his inaugural address that he had 'no lawful right' and 'no inclination' to interfere with slavery in the South." Certainly racial ill-feeling was virulent in the North before, during, and after the Civil War despite the escalation of antislavery ideology for war purposes. During the war, the Union Army carried out studies in order to assess the plans of Negroes with respect to government out of the South. Indeed, the fear of inundation by millions of "semi-savages" led Illinois to endorse new Negro exclusion measures in a referendum held less than a year before the Emancipation Proclamation was issued. In defense of the Proclamation, Lincoln himself asserted that blacks were most likely to remain in the South. "There is no reason to believe," V. J. Voegeli, another revisionist student of the period, asserts, "that full equality for Negroes ever became one of their war aims." In the Northeast and the Midwest there continued to be a great deal of bold talk but pitifully little related action. Negroes remained the victims of social discrimination, social ostracism, and injustice firmly secured in economic subordination.[15]

This is the American dilemma which Myrdal pinpointed. But the "dilemma" is much more broadly rooted in the cultural and institutional infrastructure of the nation than its limited use with respect to the American Negro has traditionally suggested. Why, after all, is the nation ruled by a two-party system in which party differences are at best obscure? Neither the extreme left, including the theoretical liberal, nor the reactionary right, both of which observe decay in the social fabric, are regularly looked to for resolution. Rather the assumption of the society's essential health, which marks both the concessionary liberal and the moderate conservative, makes them partners in the workaday resolution of the overriding dilemma of severe social inequalities in America which blacks suffer in the extreme but not exclusively.

The racism which accompanied frontier abolitionism makes us question Turner's assumption that social conflict in America was usually resolved by democratization, the keystone of his commitment to the "frontier thesis," which argued, incidentally, that with the closing of the nation's physical frontier, a new technological frontier would fill the vacuum—the seeds of that technological development were already present in the early twentieth century for Turner to observe—and the public school would be an indisputable part of it. Turner was, of course, finding historical justification for the society he was committed to perceiving as a reality—a democratic reality which was in fact defined in terms of large numbers at all times being excluded from its benefits. Those excluded might change—if they were black they might not—but the acceptability of suffering is implicit in all such observations and rationalizations.[16]

Abolitionists knew that white supremacy must remain unchallenged to disestablish slavery. That meant piecemeal reform, finally it meant piecemeal reform at horrendous cost. But the fact remains that such piecemeal reform did nothing and has done nothing to break down its own prerequisite—the necessity of seeing Americans for all purposes, as two quite separate groups, black and white, a view which permeates the whole country and was, from quite early on as we shall see, reflected and reinforced in the public school.[17]

It is important to understand that the "limited" attempts to confront the question of race in America have been of necessity limited by the nation's refusal to face wider questions of social and economic inequality that might seriously threaten American definitions of well-being, growth, and progress. These definitions contain racism and make it possible—even necessary —but they do not explain it. The special nature and historic constancy of black problems provide a privileged point of entry into an understanding of the system as a whole. The long-standing plight of blacks in America and the degree to which they have come to dominate descriptions of social decay and urban strife are a blatant reminder of the gap which separates the American belief that all social problems are solvable through the school and the fact that many problems remain unsolved for generation after generation because the very assumptions on which institutionalized democratic rhetoric is built act to pre-serve, not to transform, the social order.

Both the growth of the common school and the movement for the abolition of slavery are representative of and intrinsic to the formation of what has been called a new society—urban, industrial society. What we tend to forget is that while urban society does entail a change in means of production, consumption, and distribution, it represents only a redefinition of social terms and no radical change—no transformation—in the de-pendence of power and success on their perennial corollaries, powerlessness and failure. Modernization should not be con-fused with social progress.[18]

Slavery was not at all a serviceable institution to a developing consumer-oriented society, but racism, the extreme example of competition and class in this society, was as basic to the smooth functioning of established class patterns as schools were to re-formulating those patterns for a modern era.

By the early nineteenth century, more children were growing up in cities than on farms; factory work replaced farming for massive numbers of workers in a relatively few years. The leap in urban population was quickly accompanied by an expansion of public schooling. And, according to most traditional views, opportunity and social mobility expanded with it.

One of the problems is that we have tended to equate the nineteenth-century term "working man" with the modern term "working class." As a result, what was primarily an institution built to serve the needs of the middle class has generally been depicted as a response to the rise of the common man. However, the works of Lee Benson and Marvin Meyers have made clear the proximity of what were once called working men to what we might more accurately refer to as aspiring men of middle-level and lower middle-level social position.[19] These were the men whose ambition still aspired to social plaudits from the most prestigious social elements as well as fusion with them. In turn the prestigious and affluent felt threatened by the changes characteristic of redefinition and sought allies among those in the vanguard of industrial growth—to protect their own cultural hegemony and resist the revolution which seemed to be tearing Europe apart.

The growth of the high school amid the more general movement to establish public schools is illustrative. High schools are, of course, one of the few places where we can get some sense of how much progress in school was dominated by class and how much it was a route to mobility for the masses. Public high schools, like public elementary schools, were presented by their promoters as mechanisms for achieving social mobility, for eliminating class distinctions—for truly democratizing society. As Michael Katz, one of the few educational historians to question traditional assumptions, shows, few poor and working-class children actually attended, and before long those who did were syphoned off into the world of work or into vocational programs "more suitable to their interests and capacities."[20]

This did not mean that elementary schools were failures, but rather that "the brazen walls of caste" were reinforced by them, not pulled down. In addition the schools provided employment for the exceptions who did rise out of the ranks of the poor and the daughters of the middle class. Containment of, or even humanitarian concern for, the poor worked then and since to expand middle-class employment possibilities in the public sector. Furthermore, public education, as Katz points out with respect to the public high school, also served the middle-class parent because "they could spread among the population at large the [fiscal] burden of educating their children." Professional, clerical, and unskilled manpower could be prepared for a single cost, replacing the separate expenses of charity and private tuition.[21]

The traditional view of the triumph of the common man over conflicting and more powerful interests has lent weight to another distortion. Namely, it has supported the view that those who opposed Horace Mann and his allies in other states were opposed to education. They were, in fact, arguing for a different sort of education. They had not yet recognized the irrelevance in a new society of old accrediting patterns and their curricular bases. Mann's common school was conceived as part of an emerging social order to be governed by a public philosophy requiring more state intervention. Mann's critics sensed that the new professionalization and centralization would interfere with community interests and lead toward too much state interference in the life of the individual. And, of course, they were right.

The school reformers, in opposition to this "conservative" view, worked successfully to establish the elements of a standard system: graded schools, formal teacher-training, centralized administration. But equally successfully—despite their stated concerns for "natural" non-coercive methods of instruction and for creating citizens for township democracy—what they spoke about quickly became a series of bureaucracies supportive of the very same social immobility and inhumaneness that existed in the world beyond the school.

For example, let us look at the graded classroom, an important nineteenth-century school innovation. The graded classroom was secured in principle before the Civil War and increasingly implemented after the war. It was designed to make it possible for the teacher, by keeping children of the same ability in one classroom, to teach according to their specific intellectual needs rather than to a lowest common denominator for a class made up of varying ages and abilities. Whatever the rhetoric of increased individual opportunities, there is reason to believe— although we have not believed it somehow—that the selection was then as now a reflection of social class. That this was not entirely lost on contemporaries is suggested by their realization that graded schools reduced the problems of disciplining and organizing unwilling and less able students—the inevitable other side of the coin of selection for class. Mann's belief that "Few things can have a worse effect upon a child's character than to set down a row of black marks against him, at the end of every lesson" was quite neglected by the growth of a system which enshrined black marks in class organization and rewarded (or punished) student effort by the systematic progression through grades. To be "left back" theoretically acknowledged no more than lack of effort.[22] But that was not all it was; it was a self-fulfilling row of black marks. By the end of the century William Torrey Harris (U.S. Commissioner of Education at the turn of the century) could refer to the graded school as "perfecting the habit of moving in concert with others" and as a moral force "training in the social habits . . . regularity, punctuality, orderly concerted action and self restraint."[23] New terms, not new relationships, the precarious provinciality of institutions, the conviction of stability of forms—which Thoreau observed as "warped and narrowed by an exclusive devotion to trade and commerce and manufactures and agriculture and the like, which are but means, and not the end"—dominated educational reality and sustained its illusions.[24]

Similar reasons were predominant when it became important for truants to stay in school and for willing pupils to stay in school longer. The child-centered approach, which has increasingly characterized the liberal commitment to more schooling since its early formulation by Horace Mann, grew less out of any abhorrence for the social relations of the classroom than as a shorthand statement of the need to reach the child, retain him, socialize him through an efficient pedagogy. The so-called progressive education movement was inspired not so much by a child-centered sentiment (although some individuals in it probably were), but by the growing need to define school efficacy in terms of a much more comprehensive and longer-lasting custodial function.[25]

My purpose is not to deny the sincerity of Mann's query, "Would the trainer of horses deserve any compensation, or have any custom, if the first draughts which he should impose upon the young animals were beyond their ability to move?" But this "soft line," as Mann's critics considered it, became not "soft" at all but the core of a new relevance, as restrictive as the old. Compassion and sophistication were the operating rhetoric on the basis of which the absolute numbers in the middle class were expanded but the relative proportion of poor people to the nation as a whole remained much the same. Despite the best hopes of its pioneers, the public school is an expression of the consensus which created and maintains it. As is so often the case, it was a conservative perception, expressed in 1850, which grasped quite clearly how difficult real social change would be. As, in fact, it is. In a nation of millions of disparate stations in life and engaged in a large variety of pursuits, a conservative considered it empty rhetoric to talk of equality in and through education. "A scheme of universal equal education, attempted in reality . . . could not be used with any degree of equality of profit."[26]

Why bother then? The point is, nobody really did. There was no evidence of movement in the direction of truly "equal education"—or, as we call it more recently, "equal educational opportunity"—and the best critics of early American schooling saw its authoritarian nature and class discriminatory intent right away. Ralph Emerson feared for the survival of individual selfhood in the new public schools. David Crockett feared that elitists would deny real educational opportunities to the less fortunate since the better part of land and tax allocations went to higher education, not to grade-school education. Both believed the schools were spoiled from birth, given the cultural and social pressures on them.[27]

The common school's mission was to maintain and transmit the values considered necessary to prevent political, social, or economic upheaval. Daniel Webster said public education is a " . . . wise and liberal system of police by which property, and life, and the peace of society are secured." Horace Mann viewed the school's chief function as a social and economic "balance wheel" balancing conflicting interests and thus protecting the social order. The common school would protect the rights of property, first, by teaching the children of the propertyless to believe that the economic system was reasonable and just, rewarding to people according to their natural abilities and real contribution to society; and second, by teaching that if one practiced those Puritan virtues, he too could be successful. Mann was very clear about the social impact such a school might have on revolutionaries:

Finally, in regard to those who possess the largest shares in the stock of worldly goods, could there, in your opinion, be any police so vigilant and effective, for the protection of all rights of person, property and character, as such a sound and comprehensive education and training as our system of common schools could be made to impart; and would not the payment of a sufficient tax to make such education and training universal, be the cheapest means of self-protection and insurance?

Cheap or not, it worked. There were no violent revolutions in America with the exception perhaps of the Civil War.[28]

America—alone of major Western societies—produced an educational ideology to match its needs and fears in the wake of massive industrial expansion and huge democratic upheaval. Nothing is so familiar in both popular and scholarly images of America's open society as the public school. This was the great practical and intellectual achievement running parallel to and supporting industrialization; largely it is a creation of the middle class in the nineteenth and the beginning of the twentieth centuries. As has been suggested, it was the birth of a new formal system of assimilation into the social order, replacing the church and its various charity facilities as a mechanism for reinforcing certain kinds of familial patterns and punishing others. The school did not replace the family; it simply interpreted traditional patterns of familial dominance into an increasingly urban, industrial setting with its new manpower and consumer demands. It also failed in any serious way to modify the lot of the poor. It promised a way out of poverty for those who could use it—the society has never been willing to tolerate more than that.

The most obvious political characteristic of the nineteenth- and twentieth-century public school is that it lives by and transmits the values and attitudes appropriate to its own dominant middle-class life-style—a style that emerged from an older middle class, from the captains of agrarian forms of industry, and which was acceded to and refurbished by those rising through the wealth derived from urban, industrial enterprise. In Europe groups rising through industry could only make headway by ideological embroilment with the established governing class. Studies show that time and time again nineteenth-century European revolutions—1830 and again in 1848—were inspired and taken over by middle-level men restricted by the established order. There was no such ideological revolution in America or in Britain, hence the peaceful nature of the so-called revolutions and the rise and consolidation of an industrial elite and middle class in those countries. In America, the public school reflected the accommodation.[29]

The one creed of the middle class, produced by its intellectuals and epitomized by the public school, was utilitarianism, which was little more than a justification of the social system. It was crippled by its perennial refusal to allow morality and ethics into the practical world. What was practical was a reality orientation; what was moral lived only in the realm of philosophy or in a world yet to come. Since morality had been removed from the practical spheres of human life, it became meaningless except as an expression of human hope. The school, like the church, maintained this colossal contradiction, subordinating human development to the demands for maximal output.

The ruling ideology of this social organization was a combination of "traditionalism" and "empiricism," intensely hierarchical, which reiterated the strength of the dominant agrarian class to which the bourgeoisie assented and which it mimicked. By the end of the nineteenth century the school reflected the firm establishment of a new, broader class, "common" to the nation as a whole outside of the South. The new class never had to rethink its relation to society to gain a foothold on its comfortable social station; society was therefore reinforced as an immutable natural order. The driving force of the American economy made this unity possible. With no feudal origins, American agrarian enterprise was capitalist early and so no overthrow of a previous ruling class was necessary to achieve a different common mode of production. The fusion of the new class and its role within the existing social structure was consolidated still further by the phenomenon of underpopulation which required America to impart labor in order to support continual expansion. The class issue was rapidly formulated in racial and in ethnic terms, and by the end of the nineteenth century the school stood for Anglo-Saxon dominance.[30]

Once the new class had secured itself, it was resolutely hostile to any form of thought which put the whole social system in question. It organized itself against such potential subversion. Slavery might be abolished if the costs to social order were not too high; the poor should be fed since their marginality was important to the system and their illegal search for sustenance contaminated the urban milieu. The public school emerged to make it possible to contain and reach the undesirable behavior of the mass urban poor while at the same time reducing the high costs of maintaining both private charity schools for the poor and private schooling for the privileged.

There was no 1848 revolution in America, only the fight for and eventual consensus toward the establishment of a public school system, which by 1860 had begun to appear in a majority of the states. Between 1800 and 1900 the public school victory was consolidated and in process of new redefinition for new immigrants, new industrial expansion, and new cities. Once again the "soft" line—child-centered approaches, individualized instruction, teacher-training—was the theme of reform, and with the aid of settlement houses and other charitable agencies the public responsibility for social problems was expanded yet again. A public sector—public health in schools, free lunches, welfare workers—was being firmly established around the school. The progressive ideal, the "community school," through increased professionalization and credentialism comprised a shorthand statement for what has become known as a "middle-class welfare establishment" offering few and negligible gains to the masses of poor people out of whose need the web of humanitarian rhetoric and public conscience is spun. The poor themselves have been subjected to oppression by these new groups in much the manner H. G. Wells recognized in his *New Machiavellians*. Reformers and their workaday agents have been concerned with defining the territory of their own prerogatives and security while aiming to ameliorate the condition of the less than fortunate members of society by demanding that they raise their moral standards as evidenced by hard work, temperance, and providence. "Thou Shalt Not" was equally and simultaneously the watchword of social reform and social control.

In 1900 the secular bourgeois culture was firmly secured. The harmony between the middle class and its intellectuals was complete. The old disdain of Transcendentalists like Emerson for the implicit damage formal state schools would do individuals was transformed into a schoolman's debate, heavy in philosophical considerations but specific only when it came to the best methods for maintaining discipline in the classroom. There was no separate intelligentsia publicly discussing alternate goals and methods for the schools. Critics of schools were numerous but they campaigned for the "improvement" of schools; few progressives were actually searching for a vantage point outside of the system. Social work and social reform activities were primarily concerned with the more efficient use of manpower and the reduction of waste. There was little intention and less hope of changing the system in any significant way.[31]

The teacher, according to John Dewey, bent on the "formation of the proper social life," the "maintenance of proper social order," and the "securing of the *right* social growth," would be "the prophet of the true God and the usherer in of the true kingdom of God." (Italics mine, C.G.) Dewey could disparage the dominance of the private profit motive in society—much as Jane Addams lamented the inability of teachers to understand and value the culture of ethnic minorities—but, as with Jane Addams, his stance must finally be described by his pragmatism which was defined in the context of society's survival. Jane Addams, for example, was anxious for mobility for immigrants, at least so far as expression of Americanization might imply it, but finally she came down on the side of stasis within the security of an active program of Americanization; to be good parents, she argued, people must first be good children. Allegiance to family was an important part of her cultural pluralism, and allegiance to family rooted in peasant and European tradition meant quite the opposite of individual progress in most cases. It meant ethnic-class progress at best, and relative group immobility very often.[32]

Further, Dewey believed that the good pragmatist should not waste his "influence" on lost causes. In the name of survival and effectiveness, utilitarianism countenances compromise. Survival was the cornerstone of economic rationality—of supervision and controllability. The school was necessary ballast, especially as Dewey saw it, for the "expert manipulation of men in masses for ends not clearly seen by them, but which they are led to believe are of great importance for them."[33]

The kind of progressivism implied in this view is not at all the progressivism W.A.C. Stewart had in mind when he defined it in the context of radical social change, namely a series of reactions to schooling that has become rigid, in order to liberate what has become repressed.[34] Its allegiance is clearly much closer to the oppressor than to the oppressed. The links which historians have frequently observed between American educational and political progressivism after the turn of the twentieth century are real enough, but they hardly represent a new high point in attempts to resolve social inequalities. So much for middle-class transformations.

5

The Assimilation of the Immigrants: The School Didn't Do It

By the beginning of the twentieth century, urban public schools were already considered part of the fixed order of things. They were believed to provide common experience for the diverse children they were to equip with the wherewithal for responsible citizenship and economic prosperity. Despite a rhetorical concern for the underdog, schooling in practice remained a highly selective process, expanding its base to meet changes in the world outside, never to provide the means by which less than privileged youngsters might gather the wherewithal to act on society. "Americanization" always meant making "the other races" accept the promises and style of the dominant Protestant culture with the clearest white image of itself. In turn, the widespread economic success of immigrants assured the sanctity of the school in the American democratic experiment.

The ruling elite's narrow definition of "Americanization" required making the right "socializing" influences accessible to foreigners. It also entailed guarding the dominant culture and those privileged by it from such undesirable influences as Catholic priests and party bosses. The school reform movement stood solidly in defense of the white Anglo-Saxon Protestant dominance. New York's Blaine Amendment, and other state legislation like it in most Northeastern and Midwestern states, which bans the use of tax funds to support church-governed institutions, was much more than one more piece of legislation in the history of constitutional checks against church interference. Both the Blaine Amendent and the almost simultaneous consolidation of public school authority in New York City were attempts to impose desegregation on the officers, professionals, politicians, and students in the autonomous local school districts which existed before 1900.[1]

The actual intention of both the Blaine Amendment and the drive toward centralization is evident in the blunt words of none other than educational historian Ellwood Cubberley, on whose work, as I have noted, so much school mythology rests. Cubberley thought that schools should have the legal right "to so classify their schools as to separate those who are overaged, defective, delinquent, or of the Negro race."[2]

Clearly, it was the local immigrant politicians who were the threat to be defended against. Blaine put the burden of desegregation on schools and the increased numbers which resulted from even its piecemeal execution led university professors, leading schoolmen (who often exchanged jobs during their careers), and social reformers—almost all of whom were WASPs—to move toward centralization to constrain local "politicians." Their purpose was to put control of urban schools (of immigrant schoolchildren) in boards which were small rather than large, unrepresentative rather than unwieldy and inefficient, and elected from notables in the city as a whole rather than by wards and the more broadly based "partisan politics" wards represented.[3]

There were valid questions of graft and ineptitude raised against large school boards and decentralized power, but mostly the problem was the uncertain power of conventional sources of authority. This became clearer after the successful consolidation of city school systems. Instead of dominant Protestant group control there was rapidly a transfer of local ethnic political pressures to the central administration. With each decade "the gentleman in public office" lost his place on politically appointed

Boards of Education to Irishmen, Jews, and by the late 1930s, some Italians too. In their attempt to continue to influence school policy the old elite retreated to extra-school agencies such as the Public Education Association, the National Education Association, the Settlement House Movement, the university schools of education, to exercise power on a system from which their very places in the establishment precluded them. The reform platform they adopted expressed both their concern for the circumstances of urban poverty as well as their concern over their increasing distance from the institutional centers of urban electoral power. Their alliance with the victor was generated in part, at least, out of a desire to protect society from the least desirable groups while at the same time expanding the routes to advancement, without jeopardizing social equilibrium, for those groups deemed worthy of competition for mobility.[4]

The theme of immigrant assimilation in the United States has been indispensable to the "Great School Legend." According to the conventional wisdom about immigrant assimilation, it was the public school which made America's unique adventure with ethnic diversity possible. The belief that the vast majority of immigrants to the United States since the late nineteenth century have been successfully included in the nation's dominant middle-class affluence has meant that the existing problem of black poverty in inner cities is explained by the experience of slavery or the deprivations of urban poverty itself. If blacks don't yet make it and it isn't their fault exactly, it is—in the optimism of our most enlightened policymakers and the social scientists who serve them—the fault of something in Negro history made manifest in the public school.[5]

As Nathan Glazer and Daniel Patrick Moynihan put it in *Beyond the Melting Pot:*

There is little question where the major part of the answer must be found: in the home and family and community—not in its overt values, which as we have seen are positive in relation to education, but in its conditions and circumstances. It is there that the heritage of two hundred years of slavery and a hundred years of discrimination is concentrated; and it is here that we find the serious obstacles to the ability to make use of a free educational system to advance into higher occupations and to eliminate the massive social problems that afflict colored Americans and the city.

The big lie as normative reality! With the public school historic and strong, city problems, and problems in city schools, are black problems now not because blacks have them but because blacks, alone, fail to conquer them.[6]

But the story of "immigrant" success is a legend too. Some groups did better than others, and some parts of some groups did best of all. But it has been equally true that some groups did less well than others, and some parts of some groups did worst of all. And over and over again it is clear that even the school's ability to retain pupils coincides with and seems to be dominated by that group's simultaneously recorded adult employment rate. We have been bemused by the successes—the Jews in particular—and the easy location of failure—among blacks for example. But many millions in all ethnic groupings have suffered miserably in America and continue to do so.

The high degree of school achievement preceding disproportionate economic success among Jews, for example, which has confirmed our expectation of public schools, did not mean success for all Jews. Otherwise, why the remedial classes and dropout panic in several of the schools on New York's Lower East Side with as much as 99 percent "Hebrew" registration? Where the family was poor enough to take in boarders to cover rental costs, and desperate enough to join the city's welfare rolls, delinquency and criminality were then, as they are now in some urban neighborhoods, the burden of Jewish families too.[7]

There is a hard core of reality behind the story which depicts the entry of the Eastern European Jewish immigrant into the small business enterprise and then of his son into the university and the professions. The "business" quality of the ethnic community has not itself been the vital ingredient; the key factor is more probably the indigenous grounding of the unit within the ethnic boundary, that is, the establishment of an ethnic middle-class before scaling the walls of the dominant society.[8]

Economic stability for the group preceded its entry onto the

broader middle-class stage via education. The correlation between school achievement and economic status was so high that
in school surveys carried out in the Midwest during the 1920s
it became necessary to separate Scandinavian-Americans from
other "ethnic" Americans because the school performance of
their children so outdistanced other foreign Midwest groups.
Census figures reveal that the degree of economic security
among farm-holding Scandinavians and storekeeping Jews (surprisingly high even in 1920) was much greater than among
more characteristically wage-laboring groups. As Theodore
Saloutos points out, the Greeks, too, were among the first to
make this transition. Similarly, in 1940 the relatively recently
arrived San Francisco Japanese community—tightly organized
around their own business enterprise—ranked high on school
achievement measures.[9]

But Oscar Handlin's presentation of an "ideal type" immigrant in his seminal work *The Uprooted* once established as the
orthodoxy in the field has proved difficult to dislodge.[10] Rudolph
Vecoli showed for Italians what Thomas and Znaniecki had
shown for Poles in 1929—that the occupational experience of
the homeland was a significant ingredient in the pattern of adaptation in the United States, and therefore that immigrant experiences were both turbulent and varied.[11] Moses Rischin recently
made it clear that the outstanding rate of Jewish mobility in
the cities of the Northern United States depended in large measure on their urban and small township entrepreneurial experience in Eastern Europe.[12]

Each of these works also pictures the living conditions of
lower-class workers in New York and other major cities in the
early decades of the twentieth century. A large body of contemporary observations made the miseries of factory life staggeringly clear. Isaac Hourwich showed in 1912 that even the
more favored "old" immigrant from Western and Northern
Europe progressed at much slower rates and in much lower proportions than was contended by those who opposed "new" immigration from Southern and Eastern Europe.[13] Robert E.
Park, Herbert A. Miller, Jacob Riis, Jane Addams, Lillian

Wald, and many others confirmed the picture with moving descriptions of what they saw among the urban lower classes.[14] United States census data for 1910, 1920, and 1930 reveal that assumptions of school success preceding social progress where it occurred are as ill-informed as most popular assumptions about the inevitability of mobility itself.[15] Things seem to have worked in quite the reverse order, with cultural background and economic status being reflected and reinforced in the school, not caused by it. In that sense, the school was a less effective route up the social and economic ladder than we have generally believed. John J. Kane, a sociologist at Notre Dame, put the problem this way with respect to the Irish:

There may be some kind of lower middle or lower class orientation among them to education and occupation which tends to anchor Catholics in the lower socioeconomic groups and which limits those who do achieve higher education to certain fields which appear to offer more security, albeit less prestige and income. It may also be that leadership, even outside the purely religious field, is still considered a clerical prerogative, and the same seems equally true of scholarship. It seems that Catholics creep forward rather than stride forward in American society and the position of American Catholics in the mid-twentieth century is better, but not so much better than it was a century ago. Neither is it as high as one might expect from such a sizeable minority with a large educational system and reputed equality of opportunity in a democracy.[16]

From very early on, schools seem to have had highly differential effects among the ethnic groups. Immigration Commission research workers found in 1911 that even controlling for exposure to America—length of residence—ethnic differences still predominated, leaving Irishmen and Italians considerably less advanced than Russian Jews.[17] This finding was confirmed between 1911 and 1920 in big cities like New York, smaller cities like St. Paul and Minneapolis in Minnesota, and confirmed yet again by smaller studies of expanding towns such as Hartford, Connecticut.[18] Census data in 1920 and in succeeding decades up to and including 1960 made it clear that even when immigrants became Americans, neither schools nor society offered quite the mobility imagined.[19]

In 1950 more than 80% of New York and New Jersey's working men of Italian, Irish, and Slavic extraction were employed in unskilled or semiskilled occupations. Twenty years later the same phenomenon persisted nationwide. What have now become known as "white ethnic groups" continued to drop out of school early in large numbers and continue to work in increasingly less available blue collar jobs. Indeed, analyses of census data from 1920 to 1969 show that in comparison to the sons of college educated fathers, the sons of fathers with less than eight years of education have had little effective access to college, and thus even less access to upper-level jobs. It should be noted that the persistence of this terribly underestimated social and economic immobility among the descendants of the "new immigrants" who flocked to America at the turn of the century has been a serious factor encouraging the current rediscovery of ethnicity by white working-class groups, particularly by the Italians and Slavs in Boston, New York, Baltimore, Cleveland, Pittsburgh and Detroit.[20]

What we forget is that the United States, by virtue of its simultaneous youth and rapid expansion turned to aliens for labor. Different periods brought different immigrant groups to America so that the place of respective ethnic groups in the industrial class structure was characterized by the ethnic identity of the group—an identity which to a great extent had been formed by the privileges and privations of the group in the nation of origin. But we do not generally judge Americanization by the underlying class fabric of which it was an expression. Nor do we judge ethnic identity by the American reality which preserves it in the particular ways it is preserved—what is kept, what is remembered from the homeland, and what is lost and forgotten too. Rather we judge it by the rhetoric of such expressions of democratic, egalitarian possibilities as cultural pluralism—the theoretical framework for immigrant acculturation which followed the Anglo-conformist notions of the melting pot and which bluntly stated who was to be at the top while peoples whose culture and looks came closest to the white Anglo-Saxon Protestant norm would stay furthest from the bottom.[21]

There are three major theories of assimilation. The first, Anglo-conformity—the belief on the part of "native Americans" that foreigners should give up their past cultural identity entirely and take on the social and cultural habiliments of their new homeland—is the most prevalent ideology in American history. The second theory is that of the melting pot, which presupposes a biological merger and a blending of cultures into a new American type. The third theory is that of cultural pluralism, the preservation of communal life and significant portions of the culture of an ethnic group within the context of American citizenship. It connotes political and economic integration into American society.[22]

By the end of the nineteenth century, and after many millions of immigrants had come to this country, both natives and immigrants realized that the process of mutual accommodation was much too complicated to permit the simple Anglo-conformist faith in making Americans. Politics wasn't everything, and ways of life were harder than one's political loyalty. So the notion of the melting pot became popular. Some people still meant by it that immigrants, upon arrival, ought to be melted down so that they would within a very short time resemble totally the older Americans—becoming, as it were, complete facsimiles of George Washington, Benjamin Franklin, or the passengers on the *Mayflower*. Others thought that the melting pot would produce a race of Americans different in some ways from anything that had been seen before; because different ethnic groups were entering the pot, the final product would be different. But both versions of the melting pot assumed that the final product would be homogeneous.

The trouble was that the melting pot didn't melt. Two patterns of reaction developed. Men like Yale sociologist Henry Pratt Fairchild, author of a book entitled *The Melting Pot Mistake*, concluded that many immigrant groups then entering the country could never integrate; the most that would happen was that they would lose their foreign virtues, while retaining in virulent form their foreign evils. Others, like the philosopher Horace Kallen, were more optimistic; they believed that America would be richer as the result of the persistent diversities resulting from immigrant groups that didn't melt down. Very quickly the difference in outlook became rigidified in a struggle over legislative restriction of immigrants.

Cultural pluralism has become the increasingly dominant theme since World War I, but as Milton Gordon has argued, the "ideal model of the cultural pluralist society" has never really existed. In fact, the American experience approximates some elements of this model and falls short of others. "The most salient fact," Gordon believes, "is the maintenance of the structurally separate subsocieties of the three major religions and the racial and quasi-racial groups, and even vestiges of the na onality groupings, along with a massive trend toward acculturation of all groups—particularly their native-born—to American culture patterns." He finds that "a more accurate term for the American situation is structural pluralism rather than cultural pluralism, although some of the latter also remains." In other words, "structural pluralism, then, is the major key to the understanding of the ethnic makeup of American society, while cultural pluralism is the minor one." Behavioral conformity is achieved but not structural integration in many cases. The great majority of newcomers and their offspring have held fast to a communal life made up of their fellow-immigrants and the American community which grew up around it.[23]

Cultural pluralism did not in reality change the melting-pot dream very much. Americanization took place through the adjustment of particular peoples to the reality and limitations of the American Dream. To date we have only judged a rhetorical acculturation by a rhetorical dream. Americanization, judged on the basis of its own avowed criterion, namely equal opportunity, must be declared a charade. So, too, the belief that ethnic identity was worth preserving to provide a healthy infusion into American culture. But as a practical class framework for accommodating the multi-ethnic nature of labor in this country it is very accurate. Nathan Glazer some years ago described ethnic identities in America as "ghost-nations" deriving power from the folklore fantasy of the past handed down over generations.

He was right; but what he forgot to pinpoint was the fact of socioeconomic life which made those "ghost-nations" an *American* reality.[24]

Ruth Elson's studies of textbooks in the nineteenth century and my own study of early twentieth-century textbooks show that all minority groups, white as well as black, with the exceptions of the English, Scots, Germans, and Scandinavians were negatively portrayed. Jews, Italians, Chinese, and blacks were mean, criminal, immoral, drunken, sly, lazy, and stupid in varying degrees. But rather than this contributing to a poor self-image among the children of these groups, these texts probably are much more a symptom of the education of the more privileged children in dominant cultural attitudes toward lower-class newcomers. After all, until World War I, most poor children were not in school—they were on waiting lists due to overcrowding, on truant lists because the pressure was to work, or they were in parochial school. It was the unchanging nature of textbooks which finally allowed lower-class children to confirm their status in the world through a classroom experience dominated by the ethos characterized in textbooks; but only after the limits of ethnic desirability, the axis of American acculturation, had been established from the top down.[25]

Even the so-called rediscovery of ethnic identity in the 1920s and after, whereby settlement houses and philanthropic reform workers in general looked toward what came to be called pluralism, meant no more than acknowledgment of such superficial symbols as food and holidays. Family practices were to be homogenized as quickly as possible, morality narrowly defined; acknowledgment of native European cuisine was to be no more than a more efficient route to those ends. So stereotypes established out of paternalistic preconceptions predominated in the manifestations of so-called cultural pluralism. The case of blacks is poignantly illustrative. Here settlement workers, bent on paying deference in some small measure to ethnic background, had men acting out Uncle Remus stories in adult education classes. A harsh judgmental moral attitude, and an overriding sense of racial superiority merged in the WASP and

German Jewish social workers with an increasing professionalism in social work. The result was a strong pressure on the new immigrants to conform to established Anglo-Puritan standards. These standards formed the inspiration for the expansion of public school services.*[26]

In this context consciousness of ethnicity is a negative base, a static one, and above all, an American cultural form. Just as blacks, however their African origins were absorbed into black-American culture, are an American subgroup, and a phenomenon of the American class/culture system, so white ethnic identity is a phenomenon on that spectrum too. More groups than blacks have been led over and over again to define their ethnicity against dominant stereotypes. Proving the stereotype wrong is a major factor in what Leonard Covello, the first New York City school principal of Italian origin, referred to as Americanization by shame for those who surpassed the tortoise-like socioeconomic progress of the group, as revealed in recent scholarship.[27]

Cultural pluralism, the expression of progressive hope in ethnic diversity and national unity, is a shallow concept until pressed to the level of individual differences. Once old-world links are broken, the indigenous ethnic culture becomes increasingly characterized by modifications in response to that part of the American culture to which it is required to adjust and by the relationship of the modified ethnic culture to the total social fabric. In effect the ethnic groups, generation by generation, became aspects of an American cultural pattern—dominated by class hierarchy—not the product of a surviving projection of the old world. If the function of culture is to make for survival in a given society, then once immigration is stopped through restrictive legislation and the input of European village and township is eliminated—by the third generation at the latest—the

---

* Even by this time of the second major redefinition of public education, the private arm—the old charity base—was by no means removed. Urban programs, special classes in English, experiments in ability grouping, were funded by settlement houses and their sponsors following pilot work in the settlement houses themselves.

old world is not a given any more. When we say the extent to which children of particular origins do well or badly in school can be explained in large part by the cultural patterns of his ethnic group we are not saying that Poland prepared people better than did Italy. What we are saying is that the cultural adaptation of ethnic patterns which developed from the first meeting with America emphasized some characteristics and discouraged others. Cultural patterns in the land of origin were basic to the degree of success experienced by specific groups in America, but what the child brings with him to the public school classroom is not a pure or direct product of an historic land of origin but the combined product of some of those patterns and the patterns established by the group in order to survive in the place assigned to it in America. These are the factors which contribute to ethnic self-image and to the preparedness of children from particular groups to successfully deal with the academic and behavioral demands of the public school.

Not only has this ethnic identity been an important base for whatever mobility did exist but it has also been an Americanized clue to social status; it was, as Andrew Greeley put it, a "totemic clan" system around which the social order was organized.[28] So the quest for cultural pluralism, just as the quest for Anglo-conformity which it succeeded, confirms the class patterns in American society. The retention of ethnic identity is as important for its recognizability as for its being felt. After all, how can group diversity be meaningful in any other way when it has to exist in relation to a dominant culture which permits progress with adjustment only. The old-world patterns are gone, and what is left is a group identity which makes for solidarity and—more importantly—the imprisonment of the individual in a group ego. These groups came to America at different times and the degree of adaptation and progress each group made (each group lost its old-world ways) is now captured in an ethnic label. The cultural pluralism of which school people have been so proud since World War II is a phenomenon of class, not the successful defense of variously indigenous peoples against homogenization. On the other hand, cultural pluralism is, ironically enough, an accurate shorthand statement of what in fact did happen to immigrants and their children in public schools. Some groups, like the Jews and the Greeks, capitalized on a largely non-peasant experience in the land of origin, familiarity with urban demands, and academic learning to prosper much more as groups than did the Irish or Italians who, for the most part, had their peasant laboring backgrounds reinforced by the desirability of such skills at one point in time, thereby excluding the building of a different order of skills which would better provide security and mobility in a rapidly changing society. As school success became increasingly essential for job opportunity and as the unskilled and semi-skilled labor market continued to shrink after the Depression and the temporary priming of World War II, so those groups "permitted" to develop and strengthen the social and cultural stasis of peasant life found that their typically poor academic performance in school came to mean low status and extreme vulnerability in the marketplace. The G. I. Bill of Rights underscored these developments, but it was not until open enrollment programs were established in colleges and universities twenty years later that the children of still clearly identifiable Irish and Italian Americans began to enter higher education in significant numbers.

There was a large-scale and persistent difference in the experience of different ethnic groups after they reached America. The 1950 census revealed that the occupation levels of ethnic groups of the "old" immigration from Northern and Western Europe (England, Wales, Germany, Norway, Sweden, and Ireland) are generally higher than those of the "new" immigration from Eastern and Southern Europe (Austria, Russia, Poland, Czechoslovakia, and Italy). There are exceptions, of course. For example, of the "new immigration" the Jews in particular are at the higher end of the success continuum and the Irish of the "old immigration" are at the lower. But nevertheless success or failure could still be plotted in ethnic terms and the constancy of each group's relative place in the social hierarchy continues to suggest the essentially class nature of American ethnic designations.[29]

The heart and extent of a group's place in the social order rested on the respective abilities of various groups to meet particular conditions with appropriate culture sets and skills. It is the ethnic cultural base for such sets and skills which keeps the ethnic structure strong—indeed, essential to the class level of behavioral conformity permitted it by the dominant culture. Local stability for an ethnic group preceded its entry into the more prosperous reaches of society. The establishment of an ethnic middle-class was basic to entry onto a wider middle-class stage via public education. It was the nation's demand for manpower that set the tone for assimilation, and, as I have said before, the place of any one group on the economic ladder depended on the degree to which the culture of the former homeland coincided with the values most highly prized in the culture of the new host society. The local business, the local church and local fraternal society, followed by the factory, the union, the political machine, were agents of mobility and Americanization before the school.[30]

William Shannon, speaking about the American Irish, finds Ortega y Gasset's principle of community intactness appropriate in this context: "Groups which form a state come together and stay together for definite reasons. They do not live together in order merely to be together. They live together in order to do something together."[31] Clearly, doing things together increases the desire to be together and culture is reaffirmed out of common experience. But what it is important to understand is that it was the appropriateness of the skills and cultural background of specific groups which reinforced those traditions in a strong "hyphenated" American identity.

The years since the publication of *The Uprooted* in 1951 have not altered the correctness of the review by a distinguished Norwegian-American scholar, Professor Karen Larsen, who wrote in 1952:

Nevertheless, Professor Handlin has not fulfilled his promise. Instead of showing the effects of immigration on the 35,000,000 people who came to our shores in the nineteenth century, his book is actually a study of those immigrants only who came from the village background of central and southern Europe and were stranded in our eastern cities, notably New York. It is questionable how far the sweeping generalities of the book have a universal application, even to this group.

The point is documented by Vecoli with respect to Italians: "The historian of immigrants must study the distinctive character of each ethnic group," and not do as Handlin, who "overemphasizes the power of environment, and underestimates the toughness of cultural heritage."[32]

Rudolph Veccoli has shown how the small-township, laboring background colored the activities of Italian immigrants from southern Italy. He has also made it clear that Italian criminal organization, not at all to be denied as a vehicle for Italian security in America whatever its dubious morality, did not emerge as a response to the American city by desperate peasants, but rather was the continuation of the organized depredations of the "Black Hand," which terrorized Sicily after 1900.[33]

Irish immigrants to American cities for the most part brought with them political sensitiveness and the technique of organizing viable political associations, born out of the experience of dependence on England. Even the Molly Maguire unrest among Irish workers in Pennsylvania anthracite fields in the latter half of the nineteenth century represents a transference of similar homeland activism against English or pro-English landlords.[34]

The Scandinavians, largely small owner-farmers in Sweden and Norway, on coming to America took advantage of the availability of free soil to begin new farms. The Germans who came during the mid-nineteenth-century Scandinavian immigration did not, for the most part, benefit from free land; they more frequently located in cities, following their largely urban and township backgrounds in Germany before the large exodus promoted by the failure of the 1848 revolution. And yet the Finns, coming at the turn of the twentieth century when free soil was no longer generally available and not of very good quality, accepted deserted agricultural regions in New York and New England—abandoned farms at low prices—or "cutover" areas in Wisconsin, Michigan, and Minnesota from which the timber had been taken, because they had been farmers at home. Only a fortunate few secured good homesteads in the Dakotas and northern Minnesota.[35]

Moses Rischin has shown that Southern and Eastern European Jews were, like the Germans, largely people with urban experience. They frequently became small businessmen in America and lived primarily in cities. Attempting to a disproportionate extent to exist as small traders and peddlers, they did not always thrive economically but they invariably rose in terms of indigenous social status. In the period between 1870 and 1914, approximately 2 million Jews—or one-third of the whole Jewish population of Eastern Europe—emigrated, almost all of them to the United States and a very high percentage to New York City. Moses Rischin suggests convincingly that many Jewish men and women had experienced the life of the sweatshop well before they emigrated. When poverty intensified—in the 1800s, 6,000 Jews reputedly starved to death annually in Galicia—and after forerunners had reported from the United States that life was possible there along lines not altogether dissimiliar from those known in Europe, but with no starvation, massive emigration began. Over 66 percent of those gainfully employed in America at the turn of the century had had "industrial" experience in Europe; no other immigration approached such a percentage. Rischin describes how these immigrants played a major part in reshaping the structure of New York industry and he makes especially clear how unions and the increasing potency of strikes contributed to psychological as well as economic welfare among urban Jews.[36]

By the same token, as Robert Cross argues, if the Greeks were peasants, "they must have been peasants with a difference." Their transition in America resembles that of Eastern European Jews much more than it resembles that of largely peasant Italians, Poles, and southern Slavs. It has been suggested, although it is hard to substantiate, that they came with prior commercial experience similar to that of so large a number of Jewish immigrants, which facilitated their entrance into urban American society; it is possible that those from the countryside were neither so attached to the soil nor so involved in an extended family system as other peasant groups seem to have been. Futhermore, the Greeks appear to have come with a uniquely vivid ethnic pride. Theodore Saloutos notes, though he does not fully endorse, the theory that the most socially mobile of the Greek immigrants came from the *irridenta* where ethnic self-consciousness was particularly acute. In any event the Greeks prided themselves on their individualism. And the Greek child was encouraged by both his family and his community to make a name for himself. In significant numbers Greek-Americans did just that, flocking into business and professions. As for Jews, these were for the longest time the best avenues for Horatio Algers of either immigrant or native background. As a result, Greek life, like Jewish life has been, Saloutos claims, characterized by American middle-class values.[37]

For blacks, "integration" following the experience of slavery, symbolized the undoing of the overt segregation and separation of slave status. As such it represented a truly ethnic focus for group organization and identity, and not, as many had feared, the extermination of black ethnicity in America. The demand for integration provided a powerful rhetorical force for the insistence on achieving specific goals, such as economic security and legal equality. Both rhetoric and the concrete goals expressed real needs of the black community, and in expressing them, seems to have been a catalyst for the new and forceful black solidarity.

Booker T. Washington, now notorious for compromise, was thus in fact responding to the fact that immigrant groups were to "make it" in America by indigenous ethnic solidarity. This was what he anticipated when he spoke of blacks as one of "five fingers of the hand"—separate from other groups in things social, together with other groups in things national—in his now infamous Atlanta address. While this was not the sentimental desegregation faith we grew accustomed to after 1964, it was a realistic assessment of what American pluralism was all about. Washington responded at an economic level, believing that if blacks stayed in the South where they had roots, numbers, and a culture used to their service, they could obtain the same relationship with industrialization that immigrants had secured to the exclusion of blacks in the North. W. E. B. DuBois, who agreed with Washington in "great part" for several years after the Atlanta address, responded to the same realization, but from the point of view of a man who identified with the urban North and the world of the mind. To make room for the "talented tenth" was more important than—but did not exclude—economic advancement for the "undifferentiated mass," as he called black poor masses in general. And so he stood firm as a keen Pan-Africanist. It was not until many years later and his attachment to Communism that this position led DuBois, like a number of successful blacks, to give up on and leave America. From 1890 to 1930 DuBois' Pan-Africanism, like much of his work as editor of *Crisis*, was an effort to build a black American sense of positive identity—to make the Afro-American hyphen like any other hyphen.[38]

Despite varying white popular models of black leadership, black leaders like Washington, DuBois, even Marcus Garvey (who admired Washington) or Malcolm X, James Farmer, and Stokely Carmichael have reached for integration at the only level where it has meant anything for urban newcomers, at least since 1850, namely economic integration for the few, exclusion for the many. These leaders are a testament to the fact that blacks, like "the sons and daughters of the immigrants," (as Oscar Handlin sees immigrants but not blacks) were soon Americans; they were, according to Handlin, one of the more exposed portions of American society, and therefore particularly sensitive to its strains, but still they were soon fully a part of it. Being "a part of it" in that exclusionary context has been the overriding characteristic of minority group acculturation in this country.[39]

Assumptions that black problems are unique are now made on the basis of comparisons with a total "immigrant experience" which comes close to being today's American white society. The picture is different if intra-ethnic comparisons are made. Blau and Duncan's comparison of the mobility patterns of non-whites and whites, divided into ethnic generations—foreign-born, second-generation Americans, and native-born Americans of native parentage—suggests that it is somewhat misleading to speak crudely of "the immigrant experience." Different groups had somewhat different experiences. All made some progress of the sort depicted in popular folklore, but some made much more than others over comparable periods of time, and significant parts of all groups made minimal or no progress. While the Jews were dramatically mobile, the Irish experienced a great deal of short-term upward mobility, but unusually high downward mobility as well, and were among the slowest to rise as a group.[40]

Despite almost a half century during which there have been no mass immigrations from Europe or Asia, the fact that the picture of social mobility and social assimilation of the foreign-born and of their native-born children as being highly selective is still accurate, though no longer as trenchant and pressing a problem as it once was, means that we must not only dismiss the image of rapid mobility and assimilation, but must place, in its stead, an image of a moderately restrictive and fundamentally segregationist society. Despite the absence of an overtly structured status system on the model of post-feudal societies, issues of ethnicity, race, and culture have been superimposed on economic and occupational differences to provide a basis for discrimination, prejudice, and social inequality. The labor of millions of poverty-striken immigrants was necessary for the industrial expansion of the United States, and only because of this were its doors open to indentured servants, slaves, serfs, and, as a result, to their descendants. That fact determines the character of ethnic pluralism in this nation.[41]

Put briefly, both the term "immigrant" and the term "public school" are misleading historical generalizations. They each require specific definition in time and place—especially when being used in relation to each other—before any serious question about the relationship of schools, poverty, and assimilation can in any way be informed. The failure to do this has permitted sophisticated social scientists like Oscar Handlin to toe the line of traditional filiopietist students of immigration and public education. Looking for success in the past because they assume it in the present, they find whatever their social scientific rigor and sophistication bring to bear in their analysis of it. Tradition, the urban immigrant ghetto origins of men like Handlin and Cremin, and the drive for consensus and harmony in the late 1950s and early 1960s made immigration and the public school each both the cause and effect of American democracy.

Handlin sees blacks moving in the same direction.[42] Always displaying a great sensitivity to the suffering of the poor, white or black, Handlin manages to miss the contradiction between the object of that sympathy and his assumption about the social structure which maintains it. He misreads black history, missing entirely the terms of its horrifying consistency by insist-ing on some new dimension of suffering that is only a function of being the "last of the immigrants"—which, of course, they were not. Furthermore he misreads what in fact blacks would succeed to if indeed they were no more than successors of immigrants. This is the nub of the problem, because in strictly institutional terms—schools, hospitals, unskilled labor, inner cities—blacks have inherited the lot of earlier groups. That is to say, blacks have continued to expand the proportion of their suffering there while white lower classes have been diminishing theirs—diminishing or holding constant but not by any means disappearing.

Immigrant historians like education historians easily ignore the constancy of economic immobility and subsequent poverty in modern American history. As a result the American Negro has been viewed as though he alone suffers from the scarcity of room to breathe in the modern nation and as though the prognosis, according to the experience of other groups, were undeniably optimistic. The blacks are only the "latest" of the immigrants and will progress accordingly.

Even when writers highlight the essential difference resulting from being black in America—which, in the final analysis, separates blacks from white minority groups—they tend to support the illusion of Handlin and others that suggests that blacks will follow the same pattern in order to confirm the uniqueness of almost uninterrupted immigrant success in American society.[43]

The truth is that even under the best of circumstances, in-migration and immigration involve enormous social and psychological difficulties, whether in the movement from rural to urban areas or in the movement from one country to another. Where immigrant tradition matched dominant American values, needs, and expectations, assimilation defined by mobility took place; where matching was less sure so was progress, it seems. Where it did not exist at all there was severe deprivation generation after generation or, in some cases (largely underestimated until recently), immigrants gave up and went home or dissatisfied employers sent them home.[44]

As Marc Fried has described it, the foreign immigrant's success was more the result of a "slow, arduous, intragenerational and intergenerational change in status" than of any painless, facile assimilation.[45] The truth is that the mobility of white lower classes was never as rapid nor as sure as it has become traditional to think. The 1920 census, for example, showed that even the favored English and Welsh migrants found half their number tied to the terrifying vulnerability of unskilled labor occupations. Americans of English stock, who dominated the national language, customs, and institutions, had 40 percent of their number working in coal mines and cotton factories.[46]

The comparative approach I have criticized has been strongly reinforced by our view of the so-called Great Migration of World War I, which, occurring at a time of shrinking European immigration, brought unprecedented numbers of blacks out of the South to meet war industry manpower demands. This movement out of the South has traditionally been viewed as being both the beginning of significant Negro exposure to urban life, regardless of sizable increases in urban blacks in the decades after the Civil War, and the origin of the conditions of black ghetto life with which we are still only too well acquainted despite volumes of outrage at these conditions then and since.[47]

Thanks to the place of the Great Migration thesis in the narrative of United States history, we all know that the black population was redistributed in the 1920s, thus making the American Negroes a more characteristically urban than rural people. What we forget is the staggering resemblance of the living conditions of urban Negroes of the Great Migration period and later with the unrelenting misery in the lives of pre-World War I Negro city dwellers. These early urban blacks had, at the turn of the century, already fought for and lost the chance for decent living conditions.[48]

I do not wish to belittle the Great Migration thesis. It has given us immensely important insight into the way in which urban lower-class life breaks down, to say nothing of its having been an essential step toward different kinds of research questions. But it has also been responsible for a serious misreading of both urban Negro and immigrant urban histories. In this perspective, the use of immigrant comparisons as a base for analysis has assumed an overdependence on the post-World War I period. The fact is that this discrete temporal delineation takes no account of the increasing number of blacks who came to New York after Reconstruction to compete for industrial opportunities with foreign immigrants.

Building on Gunnar Myrdal's conclusion that "the stream of Negroes moving to the North never swelled much" before 1917, students have sought qualitative amplification of the apparently assured quantitative evidence. Oscar Handlin points out that the Negro cannot be expected to avoid the fate and patterns of mobility which applied to earlier ethnic newcomers to the city. Less optimistic observers conclude that because so many black migrants streamed into the city over so short a period of time, the Negro immigratory stream was relatively unassimilated economically, socially, and politically. On occasion, pessimistic observers of the contemporary scene try to correlate the dis-integrated family structures and social organization of the ghetto slum with the poor preparation for indigenous security and family solidarity caused by slavery. It is apparently forgotten, regardless of the date chosen for heavy Negro northward migration, that the Negro movement to the North never equaled the size of the European migration, which numbered one million immigrants per year for several years in succession. It is forgotten, too, that many an immigrant family showed all the familial shortcomings commonly attributed to a state of slavery, and that the conditions of life among urban lower classes in general produced a pattern of social adaptation appropriate to survival among dependent classes in any setting.

Throughout all of these points there appears to be a subtly different set of standards for addressing the problems of Negroes; and the difference can always be justified by the more extreme deprivation endured by blacks, whether the time is 1900 or 1970. Urban Negroes in 1900 (as was the case for those who followed) were, because of racism, confronted by very different conditions than were immigrants. The gap between Negro and European migrants as groups has steadily widened. The differences in their respective rates of social and economic progress have been intensified by changing economic demands and by the appropriateness of the tools any lower class possesses, at any given time, to meet those demands. But the remarkable thing about observations of Negro deprivation is that they have not become appreciably different over the years, nor have they become appreciably more separate from analyses of white lower-class problems since the Great Migration, although the apparent monopoly of blacks as social problems tends to suggest otherwise. From very early on it has been common practice to clothe widespread conditions of indignity and harassment in two separate uniforms, one more honorable than the other, one white and the other black. To me, the most poignant inaccuracy of urban social science has been the continuing use of a special methodology for perceiving Negro problems. Inherent in this method of perceiving, for blacks and whites alike, is the belief that no good solutions are really possible, that the public school is powerless, while simultaneously this method is used to argue quite the reverse and blame the victim. By this means blacks have been separated as a matter of concern, put outside of the broader framework within which schools were directly related to upward mobility; they are to be considered outside the general faith, and thereby the mythology of school as a regenerative agent remains intact.

One of the most remarkable features of the current debate on social inequality is that the same evidence can produce quite opposed conclusions. The considerable degree of behavioral conformity achieved by the sons of immigrants and the American success of some and not others can serve as an argument for both school powerlessness and school power. Recall that some individual immigrant groups—blacks, Mexicans, Italians, Poles—have not been included in American prosperity at the rate and speed of Jews, Greeks, and Chinese—the former through no fault of the school, the latter because of the school. Recall too that among those who did "make it" we are able to discern extremely intact ethnic group organization and identification to a degree frowned upon by school people anxious for assimilation. The integration of each within the system toward the liberty of all has little place in the unfolding reality of schooling and ethnicity in America.

And yet, who has not heard how the public school represented the fight against urban evil with its long line of victories, and who has not taken from this picture the faith that this nation is uniquely devoted to an "egalitarian" stance in a "free" society. Everything was getting better—even if that meant different things for different groups, and even if some people's better was purchased at the price of a relatively depressed standard of life for the weaker, especially the weakest, elements of society.

Indeed, the conventional view of minority group accommodation and mobility, through powerful ethnic identification, so well expressed by Nathan Glazer simply misreads and misrepresents the process.

The Negro anger is based on the fact that the system of formal equality produces so little for them. The Jewish discomfort [integration seemed to assume the breakdown of ethnic community patterns] is based on the fact that Jews discover they can no longer support the newest Negro demands . . . past cooperation loses its relevance as it dawns on Jews, and others as well, that many Negro leaders are now beginning to expect that the pattern of their advancement in American society will take quite a different form from that of the immigrant ethnic groups. This new form may well be justified by the greater sufferings that have been inflicted on the Negroes by slavery, by the loss of their traditional culture, by their deliberate exclusion from power and privilege for the past century, by the new circumstances in American society which make the old pattern of advance less effective today. But that it *is* a new form, a radically new one, for the integration of a group into American society, we must recognize.[49]

But it is not so new a form. It is no more than a different style. It is radical and threatening only to the extent that the group lowest down is among the groups wanting "in" now. Black demands do not in reality challenge the right of ethnic subcommunities to exist; that right is ideology for both whites and blacks in America. The demand for integration is, in fact, a demand organized around a quite clear ethnic base; it is built on pluralist assumptions despite its rhetorical contradiction of that viewpoint.

That this was not the case has been the consistent teaching of educational and social historians for a long time. Meanwhile, the distorted comparisons between blacks and immigrants and the traditional axioms about public schools which these comparisons support, have made it impossible to recognize the gross inequalities in American society for the descendants of immigrants as well as of slaves.

6

Immigrants and School Performance

I have suggested the essentially conservative nature of the school reform movement and the small role schools played in giving immigrants access to economic mobility and democratic cultural independence. It may now be useful to examine somewhat more closely how the public school's performance confirms this conservative aim rather than the democratic rhetoric which decorated it. Instead of acting as an agent of social mobility through improved academic preparation leading to better employment possibilities, the schools were more often agencies for maintaining social status pretty much as they found it. The school legend claims the opposite—and that is why it must now be recognized as largely mythical.

In studying a selection of American cities and townships after 1900, during a period of school accommodation to yet more "new" foreigners flooding from Europe, I found consistently that the proportion of those failing in school was always considerably higher than those who succeeded. This is not to say, of course, that those who failed in school necessarily failed in the work world, but I will suggest that those who succeeded rarely did so because of the public schools.

We do have, however, considerable evidence over the last few decades to suggest that there is a powerful relationship between school performance, social status, and employment level.

The consistent pattern of public school failure for the majority of those compelled to attend schools indicates that perhaps we are reading the success of the schools incorrectly; the successful selection of losers in this society has been as much an indicator of the school's success as the selection of winners. Excessive real mobility is a great danger to the status quo—and the public schools in America cannot be characterized by their willingness to threaten the propriety of things as they are.

Schools have been public only in the sense that what happens in them is typical of what happens outside them. For at least the last eighty years—since public school systems were typical of nationwide state organizations—socioeconomic security, as signified by employment rates and levels, has determined scholastic achievement, as measured by dropout and failure rates.

I began this study of urban school systems with the school surveys which were becoming popular as efficiency devices in schools at the turn of the twentieth century. My interest was sparked by my familiarity with a number of such survey studies in New York, Chicago, and Washington, D.C., and by my impression that they were not much different from surveys I had come across earlier in my studies of blacks in New York. These surveys were typified by either an unreasonably short duration of research time in the field or by having a well-known director, well-paid, occasional expert consultants, and truckloads of graduate students. I quickly found the consistency of the data on academic failure among so many pupils over more than seventy years more revealing of the role of the school than the consistency of survey method through an analysis of which I had hoped to ameliorate current school conditions.[1]

Surveys were stock-taking and window-dressing activities carried out to justify funds for staff expansion and new maintenance costs. Always the dominant faith was that more of the same insured invaluable improvement. School surveyists, like school historians, have engaged in preemptive analysis. As methods have become refined so the stasis of the system has been legitimized, while serious criticisms have been delegitimized through the distorting mirror of absolute numerical changes in school populations and the continuing inadequacy of funds to finance the study of new populations.

It is not enough to say, as surveyists have said for almost 100 years now, that "The institutions we have built, the practices we have refined, the provisional goals we have set as markers and guides on the way toward the good life, are inadequate to our present educational needs." The school data make it clear that even when educational needs were met they were met in accordance with social demands which permitted failure for millions. To be concerned about failure now will be no more than tinkering unless we can express that concern in the context of measuring past school failure against the concept of school adequacy. The measure of adequacy is what Frank Jennings has decribed as "reducing the possibility that any child will be defeated in his unarticulated effort to participate in the good life," but the good life as each individual is able to define it for himself. To do the old job again will answer nobody's urgency. The survey objectives have been as misleading as were the house histories.[2]

I examined the school performance data of a number of cities, but I gave particular attention to New York City, and very close attention to four others: Chicago, Philadelphia, Detroit, and Boston. My investigation was largely determined by the availability of a major periodic survey of these systems, and particularly in the period after 1900 when immigrants and their children were entering the schools in increasing numbers. Studies were made in these cities with great frequency after 1890. I have supplemented my findings with data from surveys of such smaller cities as Youngstown, Ohio; Lynn, Massachusetts; Paterson, New Jersey; and numerous others which were the subject of school study between 1910 and 1930, generally by the men who made the larger studies. And I have also used occasional surveys of other major cities—Minneapolis, Cleveland, Pittsburgh, Baltimore, Washington, D.C.—and official public school records and sectional census data where these have been available through the mid-1960s. The picture has been largely a consistent and confirmatory one, with so little difference in evidence that I am confident in making generalizations for cities as a whole in the twentieth century and even for the periods where little data was available for some of the cities I have looked at specifically.[3]

From 1890 on, so far as quantitative evidence allows us to document, the schools failed to perform up to their own claims or anywhere near the popular definition of their role. In virtually every study undertaken since that made of the Chicago schools in 1898, more children have failed in school than have succeeded, both in absolute and relative numbers. As schools expanded to match the growth of cities, so urban decay and school failure became virtually synonymous clarion calls among reformers and Jeremiahs alike.[4]

The educators and educationists who collaborated on the Chicago study and found an exceedingly high incidence of poor school performance were quick to look to foreign birth as an explanation. But immigrants were soon replaced by the native-born, and still, with each passing decade, no more than 60 percent of Chicago's public school pupils were recorded at "normal age" (grade level). The rest, to use the deceptive language of the school study, were either "overage" (one or two years behind), or "retarded" (three to five years behind). In Boston, Chicago, Detroit, Philadelphia, Pittsburgh, New York, and Minneapolis, failure rates were so high that in not one of these systems did the so-called normal group exceed 60 percent, while in several instances it fell even lower—to 49 percent in Pittsburgh, and to 35 percent in Minneapolis.[5]

The pattern of school failure has been perennially uniform, but concern for it was by no means as great as the concern on the part of educators to get more pupils into school. In 1917, and again in 1925, federal compulsory education legislation put added strength behind various state actions to this effect. The compulsory school-leaving age moved from twelve to fourteen and then to sixteen, but always with the proviso that the two years at the top were dispensable for those who either achieved a minimal grade proficiency determined by the classroom teacher or, more importantly, could prove that they had a job to go to.[6]

With the slow but certain expansion of a technological economy in the first decades of the twentieth century, the school-leaving age increased, so that the problem of caring for all grades of ability, which the elementary schools had wrestled with, now escalated to the high schools. The school-leaving age increased for those who looked forward to lucrative employment and also for those whose vulnerability as members of an expanding unskilled labor force was gradually made even more precarious by the inevitable shrinking of demand for the unskilled in a technological society, which the Great Depression underscored with terrifying certainty. Vocational instruction programs were an inevitable corollary to the academic progress and quickly became a symbol of the school's stratification role. Today, the junior college serves as the junior high school had served earlier, operating to a large extent as an extension of secondary education, with back-seat status justified by the democratic rationale of monumental numbers to be catered to.

In 1919 Chicago gave 10,000 work permits; in 1930, only 987. Between 1924 and 1930 the allocation of work permits in a number of cities was reduced by more than two-thirds. The school had not suddenly become essential to mobility, but a shrinking unskilled job market required fewer men, and so the schools were expected to help out by keeping the young in school longer.[7]

The assumption that extended schooling promotes greater academic achievement or social mobility is, however, entirely fallacious. School performance seems consistently dependent upon the socioeconomic position of the pupil's family. For example, of high school graduates who rank in the top fifth in ability among their classmates, those whose parents are in the top socioeconomic status quartile are five times more likely to enter graduate or professional schools than those of comparable ability whose parents fall in the bottom quartile. Similarly, while American males born after 1900 spend more years in school than their nineteenth-century predecessors, federal and other estimates indicate no concommitant redistribution of economic and social rewards.[8] Jews, Scandinavians, and Greeks were already practiced in the arts of self-employment, individual ambition, and the Puritan ethic with its corollary Gospel of Wealth. The truth is that the mobility of even the most favored groups was not as rapid nor as sure as it has become traditional to think. The 1920 census, for example, showed that even English and Welsh migrants found half their number tied to vulnerable unskilled labor occupations. But as I pointed out earlier, Americans of the English stock had 40 percent of their numbers at work in coal mines and cotton factories. For the Catholic peoples, the Irish and Italians, padrone and party-boss authority seemed to go hand in hand with their being classified as dull, unambitious, and generally of low intelligence by urban teachers from the earliest days of heavy immigration. Bootstraps were not classroom resources.[9]

The school failure problem was generally tucked away in xenophobic concern for expressions of loyalty and the management problems of running an "efficient" system. And efficiency was measured by the success schools enjoyed in getting youngsters *into* the classroom, almost never by the academic success or lack of it the students experienced while there. "The ratio of the number of children in school to the number in the community who ought legally to be in attendance" was the measure, and academic success was by no means a necessary concommitant.[10]

But if students were to stay in school longer, then the public school structure had to be stretched "by facilitating the progress" of those who were locked hitherto into repeating their grades. As surveyists in Chicago remarked, "vanishing opportunities of employment" meant that the time had come for "curricular offerings based on ability and purpose."[11] It was not, as Cremin suggests, "that youngsters in droves deserted the schools as irrelevant to the world of here and now" which engendered the reforms of centralization and professionalization. Most of those children had never been in school or had been in school rarely. The point was that it was becoming necessary to contain them in school now. That meant that formal institutions must expand, and have layers added to the top to allow for promotion of the hitherto unpromotable from the bottom. As in society at large, democracy here was subject to the priorities of stability and efficiency, and to control by elites—professionals.[12]

The professionals served the system and rationalized the system. Somehow the presence of increasing numbers in classrooms has reassured us of the essential spiritual integrity which Cremin finds beating in progressive hearts.

In 1931, George Strayer, with a lifetime of school evaluations behind him, looked back on progress in public education over the twenty-five preceding years. Most clear in his assessment of that progress was that not only were the top 10 percent (in terms of I.Q. scores) in high schools, but that 50 percent of all eligible students from nursery school through college were involved in formal education. That more of the nation's youngsters were in school was the point of his argument, but still more needed to be put there. He acknowledged that very high failure rates were "still characteristic" of the majority of school systems. That was not a high priority problem, however, but a job "we may certainly hope to accomplish within the next twenty-five years."[13]

The fact is that we haven't precisely because the objectives and priorities of the first twenty-five years of the twentieth century have gone unexamined and are still in effect. Paul Mort, a school surveyist sensitive to the growing alienation of urban community groups from public schools in the 1940s, blamed both the alienation and school failure rates on the historic rigidity of the system, its patent failure to consider or to plan for modification for future needs, and the fire-fighting assumptions that offered more and more of the same, making progress in education no more than the "expansion and extension of the commonplace." The capacity for future innovation and modification has been assumed because the contradiction between public school pretensions and the measure of real achievement has been entirely disregarded. Consequently, we have no valid education philosophy on which to build differently. And things are unlikely to be much different until we have first exposed our illusions, and finally addressed the problems whose symptoms we fervently wish would go away.[14]

At one time intelligence tests were considered a "measure of potential"—and this was precisely how surveyists, school supervisory personnel, and professors of education viewed I.Q. tests. It was a short step to the realization that the broadened base of high school admissions meant that academic work in the nation's high schools had to be reorganized. Very soon it was observed, too, that the amount of academic work had been considerably reduced because there were so many more students who previously had not gone beyond the fifth grade. One survey team described them as "the boys and girls of secondary age who show little promise of being able to engage profitably in the activities commonly carried out by pupils of normal or superior ability."[15]

Commitment to more and more schooling, beginning at kindergarten now (although only one in four of the eligible could go as yet) and continuing as long as possible, did nothing to modify the record of poor school performance. Compulsory attendance at higher levels only pushed failure rates into the upper grades throughout the 1920s and 1930s in such cities as Chicago, Boston, New York, Philadelphia, Detroit, and Washington, D.C.[16]

Chicago noted a 65 percent increase among the "underprivileged" between 1924 and 1931. Elementary school backwardness stood at 61.4 percent, but 41 percent of all those entering ninth grade were seriously behind, too; in tenth grade the figure was 32 percent. Apart from such factors as pupil "feeble-mindedness" as an explanation, there were school difficulties to blame, too. Overcrowding existed in Detroit, where 13,000 were in half-day sessions and 60 percent in school were "inadequately housed" in 1925;[17] in Philadelphia, Cleveland, Boston, and New York the same serious deficiencies in academic performance, the same overcrowding, unsanitary conditions, and serious financial problems prevailed throughout the 1920s.[18]

In the course of nine semesters, Philadelphia high schools lost 65 percent of incoming students at the end of the first semester, lost another 32 percent of those remaining at the end of the fourth, and were down to 19 percent of the total in the final semester. In one instance, with a 339-pupil sample established for survey purposes, only 91 survived two years.[19] Federal data on schools for the 1920s and 1930s published in 1937 showed clearly the nationwide "cumulative elimination of pupils in schools." While 1,750,000 American youngsters entered grade nine, and 86.7 percent were still in school one year later, by grade eleven only 72 percent were left, and finally 56 percent were graduated. Separate data for New York City showed just over 40 percent of ninth-grade classes graduating.[20] In the late 1940s, George Strayer recorded the same old story in Washington, D.C., New York City, and Boston. Fifty percent of Boston's ninth graders failed to graduate; in New York the figure was up to more than 55 percent.[21] In James Coleman's assessment of *Equal Educational Opportunity* in the nation (1966), in the Havighurst Study in Chicago (1964), and in the Passow Report in Washington, D.C. (1967), the narrative remains unchanged.[22]

As the number of students graduating from high school dramatically increased around and after World War II, so the function of high school has clearly been changing; as a new repository for increasing numbers of students, it became less and less an automatically accepted measure of readiness for employment. The Biennial Survey of Education of 1936–38 shows the slow but very definite growth of junior colleges: from 46 to 435 since 1917; from 1,000 to 82,000 students. It was a growth which was to become the single major vehicle for negotiating the rapidly changing relationship of a high school education to the job market. Educational requirements for most desired jobs have gone up steadily while the expansion of educational opportunities has kept pace with this development only to the extent that it has dealt with shrinking need at one level and vested interests at others through more years of school, namely through delayed selection. The increasingly inclusive nature of the educational process has not meant an increasingly inclusive opportunity picture. As the demand for unskilled labor has been shrinking the supply has been syphoned off through schooling and training but has not been redirected in any way that would balance the new circumstances of demand at these employment levels in the interests of the people traditionally victimized on the lowest rungs of the economic ladder. While conditions at these levels are not worse than they were, jobs at these levels are themselves much less certain and so, less effective in supporting faith in the American promise.[23]

Indeed the most recent national data available from the Office of Education confirms the constancy of the high degree of failure among school pupils. While blacks and minorities do worst of all, it remained true in 1970 that more than one-third of the fourteen- and fifteen-year-olds graduating from elementary school failed to complete their first year of high school. Furthermore, school retention rates among whites were still frequently differentiated by ethnic group—in the case of Italian-Americans especially. When you add the quite unsatisfactorily quantified but clearly manifest "loss" of pupils in the official record at the juncture between institutional levels—between elementary school and junior high school, between junior high school and high school, between high school and college—it becomes clear that a monumental proportion of school pupils do not experience school as the stepladder to achievement and mobility characterized in the "Legend."[24]

We have tended very often to be confused by the announcements of increased and increasing numbers of first high school, then college graduates, cited frequently to suggest the increasingly egalitarian nature of our schools. This view of higher education, and, earlier, of secondary education, as having been opened up to large numbers of those hitherto excluded misses the point that the increase is almost entirely accounted for by increased completion rates at previous levels—in high schools and elementary schools respectively—since World War I. As a result, increases in absolute numbers at higher levels are not necessarily accompanied by proportionate increases. The proportion of graduates of elementary schools, for example, who graduate from high school had not increased by 1960. In like manner, the proportion of high school graduates who graduated from college had not increased either. While the absolute numerical increase is important for those being counted, the staggering fact remains that keeping more people in school longer has not meant a significant redistribution of opportunity in the nation as a whole. This is particularly serious when one remembers how many former avenues to mobility have been closed off during the same period of time. For the most part, the oppressed and rejected remain excluded—although they are officially informed of it at higher grades in the schools.[25]

Today still, as at the turn of the century, more than 40 percent of our schoolchildren are permitted to fail. For some observers a 40 percent failure rate is a low one and an achievement for the schools. To me, it reflects all too clearly the underlying priorities of the society it serves—the society which implicitly assumes that large parts of its population must inevitably fail. It is as though society were itself a synonym for the state of war, and in any war, after all, people die.

Between the early 1900s and the Depression, the schools took credit for social mobility despite the relative immobility of millions and the at best negligible influence of the school on those for whom mobility did occur. Since, as it appeared, the schools were capable of creating mobility for some of the poor, for those who seemed "willing to take advantage of the opportunities offered them," then it seemed to follow that there was something inherently wrong with those ethnic groups which were not succeeding in school or afterward. Clearly, the fault was not to be placed with the schools. Increasingly, ethnic, genetic, and racial hypotheses were advanced to explain away the failure of some of the poor. The first victims of these theories were the Irish and the Italians; then, when poor blacks displaced poor immigrants as the majority of the urban school population, they inherited the honor of being the unteachable element in what would otherwise be an efficient and successful school system. The other ethnic poor are still in the schools and still fail, but the black failure dominates both public and professional awareness that the schools are not doing as well as they should—or as well as the "Great School Legend" claims they used to, once upon a time.

Urban public schools have failed a great many people for a long time now. In one city after another it is clear that the poor failed to make much progress in school. The failure has been, for the most part, an assumed part of school reality. Current concern for school failure is not because schools are failing to do their jobs. They are still "successful" at failing a great many young people, at screening out the poor and sending them back into the cheap labor market. But today those jobs just aren't there. The school's success at creating failure has become socially dysfunctional—and that's the source of much of the present dissatisfaction with the schools.

By 1914, the triumph of the public school as the chief mechanism of social democracy was both an article of popular faith and a canon of historical scholarship. Nevertheless, a few years later financial pressure on the schools was so great that urban school systems were actually facing a total suspension of activities. Money shortages were an annual crisis right from the start. And truancy and massive academic retardation were equally traditional and intractable public school phenomena, without seeming to diminish the claims made for, and by, the schools.[26]

New York City was typical of events in the development of urban America, a pacesetter in the formation of the modern metropolis and its public schools. Along with an exploding population came an enormous escalation of tenement facilities to house them and of public schools to do what it was believed public schools could do for people in such condition.

In 1911, Professor Paul M. Hanus of Harvard University headed a full-scale study of the New York City school system.[27] The study was completed in 1913 and recommendations were made on the basis of statistical analyses, observations of school practices, interviews with members of the teaching, supervisory, and administrative staffs. Achievement tests were used on a citywide basis for the first time. They were, however, new and not yet standardized. The Study Committee recommended greater individualization and flexibility in teaching and curricula, and confirmed the relationship between foreign background and poor school performance which Superintendent of Schools Maxwell and his various principals' committees had insisted upon since 1908. Foreign birth or parentage seemed to them to coincide with the poor home conditions which regularly accompanied high truancy rates and inadequate learning. The Hanus Survey concluded, as Maxwell had sometime earlier, that whether "overagedness" resulted from a late start in school or the simple lack of ability, the overage pupil tended to fall further and further behind. On the basis of the rather limited intelligence testing they did, the survey concluded that special assistance was needed for about 15,000 children, while 2 percent of the children could be classified as mentally retarded. Despite the Hanus conclusions, the Maxwell promptings, and the scattered evening classes for adult immigrants throughout the city, no real immigrant education program was shaped by the Board of Education until the fear of alien influences during World War I and the Red Scare which followed it led to frantic defensive action. The schools were left almost alone to help the immigrants negotiate a way through their difficulties with some assistance from settlement houses, until xenophobia declared—for the short duration of the war—that the fact that schools had failed to help foreigners was recognizable—and possibly dangerous. But this was a brief period of doubt. The faith in the efficacy of the schools was quickly restored after 1920.

Who *were* the children in the schools during the period when the Hanus and Maxwell studies found the city so unsatisfactory? The census of 1900 showed 1,250,000 of New York City's 3,500,000 inhabitants were foreign-born. The foreign-born and their children constituted 85 percent of the public school enrollment. In 1910, immigrants and their children were still in excess of 70 percent of the school population. When Grace Abbott reviewed the proportions in 1917, the figure for this group of children remained above 70 percent.[28]

But if school enrollment swelled under the stimulus of immigration, the system did not expand accordingly, so that, as
Maxwell saw it, an "antiquated system" made hundreds of
thousands of pupils "of all races and conditions" subject to in-
sufficient programs and inadequate facilities. The schools, idealized to provide the Americanizing influence necessary to make
the melting pot fuse its ingredients, to act as "the great equalizer
of the conditions of men," instead actually confronted the
lower-class immigrant child with a school setting indistinguish-
able from the poverty he knew so well at home. Both Superintendent Maxwell and Mayor Van Wyck regularly stressed
that the lack of funds available to education made it impossible
for the schools to meet their growing commitments. Overcrowding, half-time classes or part-time classes, and truancy were
the rule "in nearly all sections of the city." Overcrowded school-
rooms were not merely tightly packed but stank as well. Jacob
Riis was disgusted by the immigrant schools he visited. They
were ill lit and ill ventilated, overcrowded, and rat infested.
Joseph Rice, the school muckraker, made it clear that New
York was among the best of a very bad lot in expanding urban
education. At the turn of the century 1,100 willing children
were refused admission to any school for lack of space. Pupils
in part-time classes had increased from almost 10,000 to 45,000
since 1898, and in 1902 they numbered 70,000; by 1906,
over 14 percent of the elementary school registers were on
such sessions with the heaviest strain on first-grade classes.[29]

On the Lower East Side (P.S. 137) and in Brooklyn (P.S.
84), classrooms were drastically overcrowded and the schools
themselves 10 percent overpopulated. In the congested business
area of the city, smoke penetrated the crowded classrooms and,
with the few exceptions provided in more favorable parts of the
city, the heat generated by cramped conditions and city business (76 degrees in February at one Brooklyn school) was a
constant "menace to the health, welfare, and efficiency of
school children."[30]

The Committee on Congestion, surveying the city in 1910,

echoed and underlined the complaints described. It was the opinion of this committee that overcrowding in the classrooms had reached critical proportions. Manhattan alone was considered to be in "minimal" need of thirty-three new school buildings and recreational spaces. The Board of Education had declared a moratorium on school building in 1908 following the panic of 1907, and refused to relax it until 1910. Then the new and wholly inadequate building plan ran its course and the problems of the city schools were even worse.[31]

The school population stood at 770,243 (712,861 in elementary grades) in 1912, with a daily attendance record of 603,455. Truancy rates were high but Maxwell was not able to persuade the Board of Estimate to provide resources for pupils who did not come to school willingly. Already, education took the lion's share of departmental allocations in the city's budget and the Board of Estimate insisted that school funds be used to serve only the "willing" enrollees. Maxwell was concerned by the fact that young children, through whom alien parents might be Americanized were out in the working world as soon as it was legally possible and, more often than not, even sooner. A Department of Health report from the Lower East Side showed the high incidence of unskilled occupations—"casual work" and "blind alley occupations"—in which "children" (fourteen to eighteen years old) were engaged. Efforts to establish truant schools in the city were generally weak and ineffective.[32]

Not only was there too little room, but learning was at a surprisingly low level of achievement—according to the system's own standard of evaluation. In 1904, the first age-grade statistics for any American school system were published in New York City. Thirty-nine percent of all pupils in the elementary schools were found to be overage for their grades. The figures did not represent the phenomenon of a unique year, but a pattern well evidenced by the annual reports of succeeding superintendents. A few years later, one such superintendent's report showed that only one school in Manhattan carried less than 20 percent of its pupils overage.[33]

Superintendent Maxwell was quick to recognize that school

failure was not so much a function of intelligence as it was of
the environmental deprivation resulting from the poverty of
tenement life and the values of alien parents. Maxwell directed
attention to the fact that "the numbers of children who were
promoted from the lower half of each yearly grade is less than
those promoted from the upper half of each yearly grade," indi-
cating that the school did little to modify the ratio of success
and failure already operating among the pupils upon their
entrance into the public school. Maxwell and other educators
called for an all-out attack on the non-English-speaking home.
When he left office he found no improvement: schools simply
failed to deal with the massive numbers of newcomers; in the
previous year "we had almost the largest, if not the largest, increase in school population in the history of the city; yet no
money was granted to buy new sites or to commence new
buildings."[34]

A sample of 7,000 children in five city schools studied for the
Superintendent's Annual Report in 1907 showed that among
children in grades one through eight, more than 32 percent
were overaged. With less than 3 percent performing above age
level, the bulk of the elementary school population sampled
moved through school only until working papers could be obtained. A more broadly based school inquiry some nine years
later showed the same rates of retardation. Even these statistics
tend to give a less than accurate picture of rates because data
was computed only among those who completed their respective school years, missing the "tens on tens of thousands of
pupils" who dropped out during the school year.[35]

Paternalistically, the public school system responded to its
new foreign and lower-class clientele by catering to their "sani-
tary" and "behavioral" difficulties, rather than by attempting
to do much about academic performance. The school system
employed nurses and doctors, and developed kindergartens and
nurseries to relieve working mothers while providing models
of proper language usage and external conformity for the growing child. This, it was hoped, would make immigrant children
less visibly strangers.

Although there was a well-publicized increase in the number of high schools between 1900 and 1920, the far-reaching effects of overagedness, poor elementary school performance, and the poverty of lower-class families produced a secondary school system which was highly selective in the social background of its graduates. Of 5,410 students who graduated from high school in 1908, considerably more than half were girls. The boys had gone out to work. In 1910, the annual figures reported by the Superintendent of Schools stated that 30 percent of those who did go on to high school dropped out before the completion of their courses.[36]

Meanwhile high school entrance was itself a very selective process: in the first decades of the century more than 90 percent of the annual school population was registered at the elementary school level, while high school registers hovered between 4.7 and 6.6 percent. The urban lower class was used primarily for unskilled labor, and among this class "children set to work at the earliest age at which they can earn," and only rarely did they carry on their education beyond the legal minimum. The vast majority of the urban school body dropped out. For the urban masses, school was at best a short-term affair, and as temporary and occasional an arrangement as possible. There were, after all, more pressing needs.[37]

New York City's Department of Health found a "terrible school mortality rate from elementary to high school graduation classes" in 1911, and was concerned that children who left school to find readily available employment would be caught in a depressing cycle of future job shortage, job changes, and inadequate skills. Vocational education was the particular hobby-horse of these crusaders, but in 1911 there were still jobs available in the marketplace which could be had without specialized vocational credentials.[38]

It was evident that the city school system dealt with its clients much as they were dealt with in the wider society. Along with the complaints by school personnel and local school board officials went the realization that the city's school problems were localized but were still representative of a more general

social inequality. While P.S. 137 on Manhattan's Lower East
Side was disastrously overcrowded, and P. S. 101, like so many
other city schools, was in urgent need of ungraded classes,
there were schools in more favored parts of the city which
were running at comfortable levels of underutilization and with
reassuring rates of success in high school graduation and college or professional achievement. One investigation committee
in 1914 concluded that the public school system in New York
City has "abandoned . . . 80 percent of 14–16 year olds."[39]

When Leonard Ayres studied New York City schools at
about this time, following his less detailed studies of more than
fifty other city school systems, he showed that school failure
was clearly related to ethnic and class identification. The
records of 20,000 children from fifteen New York public elementary schools revealed that slightly more than 23 percent
of all students were at least a year behind the expected grade
for their age. The average among the entire group of cities
(about 30 percent, rising to 60 percent in individual instances)
confirmed what he had revealed in New York. Even given the
possibility of varying promotion policies, the principle of extreme selectivity within any given system is well attested.[40]

In New York Ayres pursued the relationship between nationality and retardation. His analysis reveals that retardation
was twice as great for Irish and Italian children as for students
of native or mixed parentage. Furthermore, it shows the children of German parents as being much less often retarded than
any other group, including native white Americans. The analysis
does not permit assessments of the respective weight of such
factors as ethnic differences, duration of stay, language acquisi-
tion, and social class in academic outcome. But it does tell
us that retardation was severe for some immigrant groups, mild
for others, and on the whole somewhat higher for immigrants
than native Americans.[41]

Another study of retardation in New York City secondary
schools was carried out at this time by J. K. Van Denburgh, of
Teachers College, Columbia University. Even within the highly
selected secondary school population at the turn of the century,

the results suggest a rank order of nationalities for high school completion roughly the same as that for elementary school children. The retention rates (percent of those entering high school who graduated) were .1 percent and 0 percent for Irish and Italian children, respectively; 10 percent for native whites; 10.8 percent for those from Britain; 15 percent for those from Germany; and 16 percent for Russian (mostly Jewish) children.[42]

Superintendent of Schools Ettinger reexamined the facts of school failure in 1922. Speaking to his supervisory staff, he reported that retardation among school pupils was still amazingly high, but that the incidence of school retardation was very much a geographic phenomenon. There were vast differences between boroughs and school districts, with a range running all the way from districts with well over 90 percent of school population performing according to or well above expectation for their age to schools with less than 60 percent recorded at or near grade-level "normal age." For the city as a whole, the rate of overageness and "slow progress" hovered around 46 percent, with just over 30 percent characterized as "laggards." Less than 42 percent made what could be termed "rapid" progress—which was very clearly the preserve of select populations in particular schools. In 1922, it was quite clear that in those parts of the city in which the largest number of the poor resided the highest rates of school failure would always be recorded.[43]

On the other hand, there were circumstances developing which would greatly contribute to the growing good name of the city's public school system. The public high schools, which effectively screened out the majority of public elementary school students and accepted large numbers of graduates of the city's private elementary schools, tended to concentrate the able students in relatively few—and highly visible—institutions. The success of these students was constantly brought to public attention and added an aura of quality and achievement to the whole public school system that it hardly deserved.

The true state of things in the public school system was known to its administrators and to study commissions, but still, optimism and school mythology dominated the public estimate of what the schools would do for the newcomers. Ten percent of New York City's population ten years of age and over were illiterate. Of these, approximately 90 percent were foreign-born. Although school attendance was compulsory to age fourteen, almost one-tenth of the compulsory school-age population failed to attend. A study of midtown Manhattan tenants in 1915 disclosed that a smaller-scale 1912 study was confirmed. Malnutrition and severe physical disability were widespread. In 1917, similar "serious physical defects" were discovered in another sample. In 1926, another such investigation determined that truancy, delinquency, and educational handicaps correlated heavily with the incidence of broken homes—broken as a result of death or incompatability. Only one-quarter of the truants studied came from "economically normal" homes.[44]

In 1934 Superintendent Campbell looked forward with "optimism" to the progress which was already under way in New York's educational system; overageness had been reduced, his annual report assured, but still it stood at 17 percent; together with the retardation rate which was recorded separately, "slow progress" pupils reached beyond 40 percent. While taking pride in almost imperceptible progress (his 1924 base figures did not agree with Ettinger's figures for the period), he too made the point that the "problem of overageness is most serious in a few districts and in specific schools in those districts." High school graduates continued to reflect the overall differences in school populations and the small number of successful students.[45]

With the beginning of the Depression, as shrinkage in the unskilled labor market began to insure that children would remain in schools to an older age, educators were able to include high school overageness in the accounting—due "no doubt to the fact that the older children were staying on longer because they were unable to find employment." It seemed quite clear that the Depression's toll on school building needs also depressed school performance. To contemporary observers the burden of this toll was a "district" problem and particularly in those districts which received the "very heavy foreign immigration." In forty-two city high schools, non-English-speaking parents were considered the chief reason for low rates of literacy in city school children. In 1934, for the first time on record, "the relation between reading disability and general failure" was ascertained. Three Manhattan school districts adopted this relationship as their major project. "Our reading problems date back several years"—and, as always, concern "some schools in particular."[46]

Basic to the relationship under consideration during the Depression years was the foreign element in the population. From World War I onward the unassimilated immigrant had come in for a great deal of attention. Concern for the failure of public education in the face of millions of the foreign-born had culminated in the immigration restriction legislation, but had done little to modify the partnership which persisted between lower-class status, school failure, crime, and delinquency. More than twenty nationalities were crowded together in the area from 42nd Street north to 106th Street, from Fifth Avenue to the East River. School districts seven and nine which served this area typified the school problems of lower-class populations.

In the areas above 106th Street, it was found that the "diversity of races and peoples" in these districts, living at pitifully low levels of subsistence, continued to sustain meager rates of school progress. J. B. Maller's study of retardation concluded that "there are many schools in the city in which the rate of progress is many times higher than in other schools," but that the striking differences "are due only to a small extent to differences in school practices, in principals, and in teachers . . . [but] chiefly if not entirely to differences in the nature of the pupils enrolled. . . ."[47]

So the old pattern of school failure for poor people continued. It continued, in fact, in the period when, according to traditional histories, school adaptation to large-scale immigration and the first clear evidences of its melting-pot magic are assumed to have been successfully demonstrated.

Studies of school progress during the 1930s were very precise about the location of failure in New York's school system. Approximately 30 percent of the school population was classified as slow learners, but the distribution of this failure was again clearly related to the social class of the students. In some schools, for example, "slow progress" was maintained by less than 5 percent of the pupils, while in others the rate was in excess of 50 percent. Similarly, in some schools "rapid progress" was as high as 35 percent of the school population; in others it was less than 3 percent.[48] Using federal 1930 census data for 310 "health area" divisions in New York City, Dr. Maller, of the Lincoln School, studied 270 neighborhoods with a population of 26,000, each of which had two elementary schools. Out of this data and measures of intelligence, he was able to correlate school achievement significantly with I.Q. and with such factors as annual expenditure per family, birth rate, death rate, infant mortality rate, and finally with ethnicity. Quoting a similar study of 1922, Maller proceeds to confirm its findings, but with interesting differences. In 1931, "slow progress" in school was found to be lowest among Jews. In 1922, this position was occupied by Protestant Americans. Italians and non-Jewish Poles were, once again, at the bottom of the scale, with only the Negroes showing slower progress. In 1930, Negroes appeared to be doing just a little better in the eleven city schools in which they predominated than they had previously, though in 1922 there was no comparable data for them.[49]

The largest ethnic blocks in the city schools were Italians and Jews, together accounting for about one-fourth of the city's school population. Although differentials in performance were accounted for by "racial differences," the studies were not referring to Negroes in reporting poor performance rates among immigrant stock in East Harlem, the Lower East Side, Lower West Side, and the Bushwick and Greenpoint areas of Brooklyn.[50]

In 1938, a survey submitted by the Welfare Council to the Greater New York Fund reported that three-quarters of the city's young men and women did not finish high school while one-tenth still did not get beyond elementary school. Nine thousand youngsters were interviewed who attended centers throughout the city maintained by private welfare agencies in whose behalf the Fund was seeking $10 million. Among those sixteen to twenty-four who were interviewed, almost 50 percent were unable to get jobs despite their willingness to work, leaving them dependent on welfare agencies.

High school populations increased considerably by the 1940s, but the ratio of success did not greatly improve. Half of the city's high school freshmen failed to graduate, and studies of dropout pupils seemed to indicate that economic pressure and the economic irrelevance of the school were the major factors in this direction. Truancy, often the first manifestation of social failure among those "deprived of most of the normal conditions for successful living," continued high and was studied as an indicator of who the subsequent dropouts would be. Prediction was more successful than any attempts at prevention.[51]

When the educational level of New York City adults (twenty-five and over) was assessed in 1943, it was found that more than 16 percent of the population had received less than five years of school—and more than 90 percent of this number still consisted of the foreign-born. By this time the legend that the public schools had Americanized and educated the foreign-born was unshakably established, yet large numbers had been untouched, for better or worse, by the schools. A little less than half of those over twenty-five in the city had only had between five and eight years of schooling. The problem of school failure was being pushed upwards. Of 6,909 pupils tested on reading ability upon entering high school, almost 90 percent scored below the norm, despite the fact that each had been awarded an elementary school diploma.[52]

Those school districts with low performance rates were those areas where "incomplete and inadequate family life . . . racial or religious problems or both . . ." contributed to existing tensions and required special educational help. "By the time these pupils have spent one year in school, their scholarship achievement is, on the average, two or three years behind that of their brothers and sisters in more favored communities." A Joint Committee on Maladjustment and Delinquency, representing the Board of Education and Board of Superintendents, concluded after two years of study (1935–1937) that a great deal had to be done by the educational system. In what sounds like an enlightened conclusion the Committee stated that "retardation is not so much a form of pupil maladjustment as it is a measure of the inadequacy of the schools." In fact, pupil inadequacy was spelled out in a new way, and it was the schools' burden to correct this inadequacy.[53]

The Depression made it imperative that more students stay in school longer. The schools had always talked about serving the poor. In fact, the children of the poor, as we have seen, had rarely stayed in school beyond the barest legal minimum, if that long. They left school to work, just as quickly as they could. But now, increasingly after 1930, the poor were "really with us" in the schools. They stayed, and for the first time, the public high schools had to face the consequences of rhetoric threatening to become a reality.[54]

As Superintendent Campbell summed it up before the Brooklyn Rotary Club in 1934, "We no longer have in our schools the selected group that we once had, and the schools as you and I knew them twenty-five years ago, would be totally inadequate to meet the demands of today. Indeed, we have a complete cross-section of society attending school at the present time."[55]

Between 1925 and 1940 three large-scale school-system surveys were carried out in New York City. The Allen Report (1924), the Graves Report (1933), and the Strayer Report (1940) continued to tell the same story. By this time, however, the problem of school failure rose as though on an escalator to the junior and senior high school levels, while all the perennial problems of funding and overcrowding followed. As the proportion of the total school population that entered high school increased, so a simultaneous decline took place in the proportions in academic high schools, together with significant increase in the percentage of those in vocational high schools (almost 50 percent from 1931 to 1939). Of 591 elementary schools in 1939, 91 were classed as "difficult" and an additional 45 as "very difficult." Considerably less than half the pupils entering ninth grade were still in school three years later, and only 30 percent actually graduated high school.[56]

Almost ten years later, in 1949, the first multifaceted remedial program—the first of what was to become a long history of such programs—to cope with "preventing and correcting maladjustment and delinquency" was established in two poor school districts; $250,000 in additional funds were to be spent in three schools over two years in what amounted to a new job for public education—a specifically directed effort to correct the effects of poverty for the black and Italian children of this area. Here, the Depression had hit harder than in many other districts because its inhabitants had not benefited from the opportunities created by "the upswings of the 1920s" as had the Lower East Side. Of the 180,000 children in this district, 60 percent were black, leaving 72,000 non-black, most of whom were the children of Italian immigrants. These poor Italian children, as the surveys had shown year after year, were doing almost as badly in school as the blacks. Yet when this pioneering remedial and compensatory program got underway, a very significant theme was sounded—one which contributed to the continued beclouding of the true history of the urban schools. The educators spoke of the program almost exclusively as though it were a program for Negroes. The failure of poor Negroes was no surprise to anyone. But the continued school failure of white immigrant children, as late as 1950, should have raised very unwelcome questions about the validity of the "Great School Legend." But these questions were not raised. Poverty, until very recently, was bundled aside as part of the Negro problem, in school and out. The failure of the white poor in school was ignored. It didn't fit the great historical model of what was supposed to have happened in the schools.[57]

7

Economic and Educational Marginality

The American tendency to see the social, economic, and educational problems of poor blacks as completely different from those of the white poor was operative even when there were relatively few blacks in major city school systems. The way in which urban white lower-class school failure at the turn of the century was explained differently from that of the parallel and simultaneous failure among urban black lower classes is perhaps even more revealing of the role of the school than that suggested by the large failure rates which existed among whites. The differences in explanation, the separate category established to examine black failure, are crucial in understanding the role of schools in American society.

In both the religious and the secular world, the Negro has long been a symbol of monumental national iniquities. Furthermore, "the Negro Problem" as a confused but blanket excuse for the widespread failure of America's dream of equality has been, as we shall see in greater detail, a device for protecting that dream. Negro failure has been used to perpetuate a "Big Lie" about non-black social progress in America.[1]

The Negro, the individual furthest down, has epitomized the inexorable relationship of success and failure, inside and outside the school. The link between permanent unemployment or chronic underemployment and educational failure is a black man now, but blacks have inherited a whirlwind no more unique to them than to lower-class whites. Employment conditions were and are most severe when it comes to the Negro, and school failure rates have been at once more glaring and more poignant. But, in effect, the public schools served Negro children just as they served the vast majority of others with low social status. This was so in Chicago, Philadelphia, Boston, and New York. That has been the problem since 1890 and it is still.[2]

But if the white lower classes have been vulnerable in the economic marketplace, the Negro, who worked sporadically and as a reserve force, was constantly a victim. If school success or failure had little meaning in the economic marketplace for whites, it bore no relevance whatever for blacks. As a result, Negro school failure was quickly isolated as a separate problem early in the twentieth century. When, for temporary (wartime) reasons in the early 1920s and in the 1940s, Negroes finally entered the lowers levels of industrial employment from which they had been excluded, those levels had already become a shrinking sector of the economy, and the massive numbers of school dropouts had no place to go. And so it remained appropriate—even inevitable—to consider Negro school performance as a separate question. But the truth is that despite the isolated and publicized success stories of particular individuals, academic effort has not often been possible for the poor and is even more rarely expected of them in reality. School studies and census data confirm the consistency of the pattern that for the most part the poor do poorly in school.

Once again New York City is illustrative of the process by which black poverty was seen as blackness, not poverty. It seemed necessary, therefore, to study it separately too. That is what the New York City school surveys did from the end of the nineteenth century until relatively recently when the assumptions of this false and misleading practice became ironically and sadly appropriate because most of the poor left in the cities were in fact black.[3]

Large-scale in-migration and wholesale foreign immigration made New York a polyglot ethnic city by 1890. While three-quarters of its population was foreign-born or of foreign parentage, a rising proportion of its population was black too—a migrant population, largely from the South. The response of educators and social reformers in general to this expanding population of blacks reveals the meanest level of the system I am examining and thereby—I believe—underscores its essential nature. As far as progressive social workers, school people, and social analysts were concerned, blacks represented a quite separate problem. They had been slower, they had yet to prove the invalidity of a historic conventional wisdom about Negro inferiority even if it was the product of an anti-democratic system, and so a different lens was fitted to the microscope of social observation, a different bacillus was identified, and an impossible miracle immunization prayed for. In order to understand the public school's dealings with blacks it is important to be clear about this general social rule despite the fact that the surrounding horrible conditions of urban life affected much larger numbers of non-black city dwellers during the first decades of the present century.

By 1914, when the Negro population of Harlem was estimated at just under 50,000 (the total Negro population in Manhattan was not much larger than 60,000 in the 1910 census), the modern Negro ghetto was established and its companion symptoms of lower-class subsistence confirmed. Harlem, however, succeeded other smaller ghetto blocks and districts which had been expanding and moving uptown since early in the nineteenth century. As it emerged and solidified during the teens of the twentieth century, it accelerated developments that had been in progress with increasing intensity since the 1890s.[4]

Negroes had lived in New York throughout the nineteenth century. By the 1850s, they had begun moving from homes in the notorious Five Points district of lower Manhattan, and by 1860 had settled in the "Little Africa" of Greenwich Village, leaving Five Points to become a predominantly Irish district. Between 1880 and 1900, the Italian immigrants moved into

the Greenwich Village neighborhoods and the Negro community made its way into what is now the midtown area. "Old
Africa," Jacob Riis observed of the Village in 1890, became
"a modern Italy."

Before the Civil War most of New York's black people had
lived below Houston Street, where the greater part of the city's
population was located. By 1900 only one-tenth of New York's
Negroes lived below 14th Street, while the number of Negroes
living north of 86th Street sextupled from 631 to 3,951 in
the twenty years between 1870 and 1890.[5]

The intra-city shift in black population was clearly reflected
in the closing of the oldest "colored schools," which began to
be eliminated soon after the Civil War for lack of pupils as the
blacks started their move uptown. By 1875, almost a decade
before municipal school desegregation laws, six colored schools
in lower Manhattan had been closed. By the time state legislation desegregated schools in the consolidated New York school
system of 1900, only one of the old colored schools remained,
and *de facto* segregation of black pupils had already been established instead in uptown Manhattan. But segregation by
ethnic groups was the rule in Five Points and Greenwich Village, too. As was poverty.[6]

Poverty dominated the lives of most New Yorkers. The Committee on Congestion, which studied New York City as a whole
in 1910, confirmed that the Lower East Side tenement was
every bit as dismaying as the Negro slum. Sudden illness or a
sudden change in the business cycle showed immediately in the
"pauper lines" of the Charity Organization Society and the City
Department of Public Charities. Among poor blacks and whites
alike supplementary income was desperately sought by ghetto
dwellers to help defray rental payments, which were always too
high in relation to income. Typically, the poor themselves
rented out rooms, creating conditions decried by reform-minded
observers because they felt the apartments were overcrowded
and without privacy. With rooms rented to boarders, the family
apartment might often be reduced to the one room that might
be used at once for "living and manufacturing." Work was

taken into the home by women who were unable to go out to work, while children worked as often and as hard as possible—their fifty cents and $1.50 per week were important to the family income.[7]

Even so, according to contemporary observers, the Negro's dependence on welfare was a much more permanent affair. Certainly his unemployment, the cause for welfare, could readily be substantiated. But amid the gross horrors of overall urban living conditions, it is interesting that it was clear to observers that Negroes were invariably housed in the worst quarters. Areas to which affluent Negroes moved were quickly surrounded by a larger population characterized by lack of opportunity.

In employment as in housing, the Negro was consistently replaced by the immigrant. Locked into the fringe employment of domestic and personal service, messengering and portering, the Negro migrant entered a society which excluded him from industrial opportunity at a time when industrialization was most expansive. Despite emancipation in New York early in the nineteenth century, the Negro had been mainly employed in the jobs for which slavery had prepared him. Between 1820 and 1860, a "job-ceiling" kept him in the homes of affluent whites as cook, coachman, and gardener; outside the personal services of the rich, he was to be found in the city doing the work of barber, caterer, waiter, and bootblack. Irish immigration gained ground in the 1840s and 1850s (by 1850 there were already more Irish servants than there were Negroes in New York City), and coincided with the Negro population's ante-bellum decline; thus it was the Irish who increasingly met the demand of the unskilled labor market and displaced the blacks.[8]

Between the Civil War and the onslaught of the "new" immigration in the 1880s, the Negro population did retain some control over one occupation, as servants who fulfilled the banqueting needs of the wealthy. The United Waiters' Beneficial Association was established in order to maintain standards and provide mutual insurance for the founding members, all of whom were black. But as French, German, and Irish help began to be substituted for blacks in the esteemed service of wealthy whites, only second-class hotels continued to engage Negroes, and the absence of Negro service became advertising fare for the menus of the better establishments. Similarly, the Negro barber lost ground in the face of foreign competition. In 1890, Jacob Riis wrote, "the Negro barber is becoming a thing of the past." Foreign-born New Yorkers by this time accounted for at least one-half of those employed in occupations with a high percentage of Negroes. As Professor Kelly Miller of Howard University put it in 1906, "the white waiter, barber, and coachman poaches defiantly upon the black man's industrial preserves." "Wide-spreading anti-Negro feeling," gripping the North with the rise in Negro population excluded the Negro from once held jobs and, as Miller stated, since 1900, the Northern Negro had been forced into "depending chiefly on odd jobs."[9]

Displacement of the Negro did not indicate a general movement of the immigrant into the service pursuits, only into preferred places; 75 percent of New York's foreign-born males earned their living outside domestic and personal service. Although the immigrant was by and large kept within the ranks of industrial labor, the practice was that he "poach" in other employment only when necessity demanded; thus surplus white labor "pressed down" on Negro jobs.

Discriminated against by trade unions, the Negro was used and characterized as a strike-breaker. He replaced strikers on the waterfronts in 1893 and 1907; on the shipping lines in 1892 and 1895; in the transit system during a subway strike in 1903; and in shipping when the teamsters struck in 1910. A nadir was reached in 1912, when the International Hotel Workers Union struck New York's hotels, and Negroes were called upon to fill the gap. In the catering trade, too, the Negro had become a "cheap man." His place on the piers and docks, even as a unionist, was as a distant second to the foreign-born invariably picked for work by Irish and Italian foremen. "The Negro gets a chance to work only when there is no one else," observed *The New Republic* in 1916.[10]

On the lowest rungs of the economic ladder, the Negro served as a relief labor force for the industrial world. He was "the last hired and the first fired," immobile in his group, and always on the fringe of urban society. *The New Republic,* convinced that the heavy European immigration was responsible for making the Negro "superfluous" in economic society, hoped that the cessation of immigration during the wartime industrial boom would be the Negro's opportunity to get a foothold in industry; rapid and increased migration out of the South would permit an end to the Negro's subservient urban status. By 1916 then, the wartime Northward migration of blacks was looked to by many observers as the chance the Negro needed to carve a secure niche for himself in the urban setting. But the coincidence of industrial opportunity and Negro migration was short-lived. The economic opportunities for unskilled labor were abruptly destroyed by the Depression, which in fact marked the end of expansion in the unskilled sector of the economy. The relationship of blacks to the economy, as a labor reserve, was still as it had always been, and they were thrown back on welfare in greater numbers than any other group.[11]

In the cities, the period from 1900 to 1914 was crucial in defining the future of the urban blacks and the role of schools. In the minds of contemporaries, consciously or unconsciously, the blacks constituted a problem different from the misery of the city and immigrant life. Somehow the Negro's past and his anomalous present united to set his circumstances entirely apart from others.[12]

The school, of course, was the place where this distortion was most clearly acted upon. To the white immigrants, the school held out the promise of eventual social mobility. But there was no pretense of seriously offering mobility to the blacks. The society had decreed the economic marginality of blacks. Their proper fate was to remain a reserve labor force, "cheap muscle" and fringe service workers. The public schools might encourage them to carry, to cook, and to clean—but there was no greater promise of full economic citizenship such as was advertised by the schools for the white immigrants. The schools agreed to carry out the mandate set by the economy. When the separate "colored schools" were closed, by 1900, the attitudes and curriculum of these schools simply moved into the new "integrated" schools. Gradually as the black population in the schools increased, the lower occupational horizon decreed for blacks became a larger part of the urban public school's expectations for all its poor students; the proper place for the poor and their children was to remain marginal workers, ever vulnerable and always available to employers on whatever terms would increase profits.[13]

As it mobilized to deal with the Negro, the education profession made very clear where it stood in relation to poverty and social change in general. Nowhere had economic marginality as direct a parallel as in the public school, and as represented there, it reveals the hypocrisy and mythology on which respect for public education rests and the relationship of its service to the economy which dominates it.

Liberal concern for the black newcomers to the city seemed real enough. The editorial which introduced the 1905 volume of *Charities* magazine, devoted to the problems of the Negro in the North, worried sincerely about the "difficulties he had still to surmount." In the face of increasing migration and immigration, the editors were concerned as to the Negro's fitness "to survive and prosper in the great Northern cities," and anxious to ascertain "how far and in what ways" the Negro migrant fulfilled Professor Boaz's anthropological claim of "genetic equality" on behalf of the black American.[14]

In 1897, a detailed analysis was undertaken by the United States government to consider the problems of the Negroes who had been making their homes in cities in increasing numbers. In 1902, Frances Kellor, an active student of urban problems and a practical contributor to government planning, published her book on *Experimental Sociology* and introduced the work with the conclusion that "the Negro child must be trained from infancy, his surroundings improved and the standards of his home life raised. Only then can the question be dealt with, 'What effect has education upon the Negro?' " She was not prepared to include the Negro in her general faith in the power of education to ameliorate the circumstances of life in the city ghettos by acting as a mechanism for upward mobility. At the bottom of the economic ladder, the Negro displayed the "typical abscesses of urban poverty," and so few individuals broke out of poverty that no pretense of the Alger myth could be seriously maintained. Clearly it had to be the Negroes' own fault—since everyone wanted so much to believe that everyone *but* the Negro was steadily advancing through the society by dint of his own efforts with the powerful help of the schools. Blame the victim.[15]

Almost as soon as increasing Negro migration to New York City was apparent, a study was conducted in New York City which set itself the task of comparing the white and Negro children in the city. In one of the "colored schools" 20 children in the primary grades, with an age range from six to fifteen years, were asked a series of questions designed to determine the extent to which the thinking of Southern Negro children was based upon the content of the Negro spiritual. Not surprisingly, the investigators found what they were looking for: the tradition of the spiritual had developed a "superstitious turn of mind." Evidence for this "Negro characteristic" was found in the children's belief that "good children who don't curse and dance and who pray" will go to heaven. Some children told the interviewers that thunder was the result of God's anger or his walking on the marble floors of heaven. With paternalistic consideration, the report concluded that the behavior of the Negro masses was not immoral, but "unmoral"—like that of ignorant and naïve children.[16]

In 1911, one of the large public schools in New York reported to the school authorities that "colored children appeared to be in need of special attention outside as well as in school." As a result of this complaint, the Public Education Association was commissioned to undertake a separate and comprehensive study of the colored school children in New York City. The study was funded anonymously and carried out within the framework of a series of studies of mentally deficient children in which the Association was engaged. The one recourse in the system available to principals with children presenting special "familial" or "behavioral" problems was the Probationary School, established by the Board of Education in 1905. By 1911, of 781 pupils from various schools in the city who had attended the Probationary School none were Negroes.[17]

In 1911 the Hanus Survey was begun. Professor Hanus reached the conclusion that the general academic level of the school population was low and that the homes of the children were serious contributing factors. Notwithstanding the general dissatisfaction with school standards, nor the existence of the Hanus Survey itself which had included black children, the school authorities commissioned a separate study of Negro school children. This study was conducted at the very time the larger study found gross citywide inadequacy. Nevertheless, those surveying black schools reported poor performance on the part of Negro children on all criteria, and placed the blame squarely on deprivation in the home, the same cause which Hanus had identified for the failure of poor children in the city as a whole. Despite the fact that Frances Blascoer, who headed the Public Education Association study, carefully documented the prejudice surrounding the Negro child from his earliest days, despite her own close experience with other ethnic groups in the city, she based her conclusions and recommendations on what she perceived as the uniquely debilitating nature of the Negro family. Environmental fatalism was replacing the belief in genetic determinism. Now the black family experience set the limits of hope. It would not be the last of such well-intentioned but misguided analyses.[18]

There were 750,000 children between the ages of six and fourteen years enrolled in the schools of New York City in 1915. Only 8,000 of these children were of Negro parentage. Manhattan's school-age population was in excess of 300,000 and the Negro school children in the city's center numbered 4,345. Despite these small proportions, the separate study of

Negroes was considered necessary. The Manhattan Negro
school population was the subject of the P.E.A. study. Most
of these children were accommodated in only nine schools in
the borough; although most heavily concentrated in some of the
uptown schools, the Negro children constituted only 4 percent
of the student body of the schools reporting to the study. P.S.
89 on Lenox Avenue, between 134th and 135th Streets, had
the largest Negro registration—1,277 in a school population of
1,841. No other school approached the Negro predominance of
this Harlem school. P.S. 119 was a girls' school at 133rd Street
and Eighth Avenue with 774 colored girls among its 2,080
pupils. P.S. 80 on West 41st Street, the last of the "colored
schools" in existence by 1900, had no colored children on its
register in 1913. The largest proportions reported in down-
town schools were at P.S. 28, 69, and 141. However, the Negro
numbers were far from large, and of 3,177 pupils in these
schools, only 493 were of Negro parentage.[19]

When a Negro child performed badly, his being Negro was
explanation enough. But, as I have shown, poor children generally performed badly, both black and white. The Negro child
was invariably poor, did perform badly, and his home left almost everything to be desired. Indeed, the sample which Blascoer selected for more detailed examination showed clearly that
school attendance and school performance improved—even
among blacks—in relation to the intactness of the family base.[20]

Hanus recommended increased spending and more adequate
resources in order that the school might do battle with the
slum and the negative influences of an immigrant home. Frances
Blascoer recommended the establishment of a "social center"
to counteract the forces in the home, because she believed that
the "statement by the school authorities that broken homes and
working mothers were largely responsible for poor attendance
and scholarship among colored children was borne out by this
investigation." Such statements were indeed substantiated. The
importance of the Blascoer study lies in the very fact that these
generally debilitating conditions were represented as a separate
problem when the victims had a black skin. And the reason

such a separation could be upheld was precisely because be-hind the similarities the problems were truly different: the Negro school dropout was not being absorbed into society—as his white counterpart was. The schools failed the white poor, but no one noticed. The white poor found jobs, which did for them what the schools claimed they would do, move them up-ward, slowly, as a result of their own efforts.[21]

Early in her introduction, Blascoer points out that the Negroes and immigrants were "different" from each other. In-deed they were, and Miss Blascoer persistently responded to the difference, noting its social and economic manifestations but preferring to make recommendations within a framework which reinforced that difference: visibility was explanation enough. As the report progressed, she added prescription to observation, recommending the development of an indigenous business enterprise in the Negro community to counteract the refusal of industry to admit the black man. How? Just do it by working at it. Forget that financing is impossible and clientele too poor. Her confusion was evident in her attack upon what she felt was the harmful dedication of the National Urban League to encouraging the training of Negro social workers; only white workers, she believed, stood "in a place of authority in relation to the schools"; only they could be of benefit to the underprivileged Negro community.[22]

The concomitant to separate economic direction was, of course, separate academic goals. "Given an education carefully adapted to his needs and a fair chance for employment, the normal child of any race will succeed, unless the burden of wrong home conditions lies too heavily upon him." Given this background in the case of the Negro, it was irresponsible, Blascoer considered, "to hold them up to an impossible stan-dard." "The outlook does not seem unpromising," she concluded hopefully with regard to occupational opportunity, if black New Yorkers would only try to develop their own indigenous economy. But first, she argued, they must forget "the factory world of the white men"—exactly the sector where immediate opportunity was being enjoyed by millions of white immigrants.[23]

Blascoer's finding seemed borne out by the most sophisticated scholarship of the period. M. J. Mayo, in his doctoral thesis for the Psychology Department of Columbia University in 1913, used the school marks of New York City high school pupils as indications of mental ability, and confirmed a long history of "scientific" opinion with respect to laggardly intellectual development among Negroes, though he left genetic assumptions aside. Using the school scores of black and white high schoolers, Mayo established that even black children who managed to get to high school performed at only "three quarters the efficiency" of whites in their studies.[24]

Mayo impresses one with his unwaveringly liberal position throughout his study. He was clearly anxious to refute the body of pseudo-scientific literature on cranial weight and brain-size measures which was popular as evidence of Negro intellectual inferiority. Having done that, he noted that any comparative racial research in the United States must be modified by an awareness of three things: first, that school marks are only relative, and by no means absolute measures of mental ability; second, that "mental ability and material prosperity will be found to be associated in a very large degree"; and third, differences in mental ability "are differences of degree," and not of kind, that "mental differences between races, as between individuals, are quantitative, not qualitative," and that "different rates of human progress . . . may be explained without the assumption of mental inequality," especially since "colored" Americans represented a "mixed not a pure" racial type. Nonetheless, he concluded on the basis of the differences he found that the findings of the physical and social sciences he had criticized at the outset had been confirmed by his research. His own findings, it turned out, were "in accord, except in their moderation, with the teachings of history and anthropology, and with the views commonly accepted among those who have made extensive observations upon the races. There seem to be no statistical grounds for holding to the view of substantial racial mental equality. Our data point clearly to a measurable degree of mental difference." His own reservations notwithstanding, he believed that this would be the view "ultimately gained from purely scientific study of the question, stripped, on the one hand, of philanthropic considerations, and on the other, of racial bias."[25]

He had apparently forgotten by this time that almost 30 percent of the colored pupils in his study reached or surpassed the median mark of the whites. The black high school population he examined remained at school longer for reasons which related to "social and economic" difficulties (vis-à-vis skilled job openings) of which Mayo was aware in this study, and of which all the principals interviewed by Blascoer gave evidence. Add to this the disinclination of all but a few blacks to contest the doors closed to them, and one is left in doubt of what exactly Mayo did prove. His findings do, of course, resemble those of Miss Blascoer; and this in essence illustrates that Mayo could not help but be a man of his time and respond to the events of the world with well-meaning but inadequate moderation. Aware of ethnic diversity as a major characteristic of New York City, Mayo was careful to include English, Germans, Irish, Italians, and Jews as well as "colored" pupils in his sample; but in interpreting his data, he saw only "whites" and "Negroes."[26]

Inevitably, Mayo touched precisely the issue Blascoer was forced to deal with, namely, that if individual Negroes were exceptions to the general citywide pattern of failure for both blacks and whites, the schools had to decide whether or not they were prepared to accept the challenge and the obviously greater expense of attempting to increase the possibility of their success. The assumption is, always, that the school can do it. Blascoer took the school off the hook, but Mayo left the question open, with a warning that "under the most favorable circumstances [the expenditure involved] would probably be on an average not less than a fourth or fifth greater per unit of colored population." In stating the challenge in terms of public school efficacy, he left the way open for increasing ineffective spending when, forty and fifty years later, the school population was largely black and something had to be done.[27]

With uncanny precision, Mayo's remarks on education parallel the society's view of the proper economic rewards to which blacks might look forward. It was not that all blacks failed, but greater proportions did. Success, while by no means common in other ethnic group experience, occurred with greater frequency in white groups—though it varied from group to group. Mayo stated that "the average mental ability of the white race [in the North] . . . is higher, but not a great deal higher, than that of the colored race [in the North]; and that as regards the matter of mental variability, the white race is more variable, but not a great deal more variable than is the Negro race." But, "In the struggle for survival, these small differences may be, and no doubt are, the determining factor." Clearly intelligence and academic achievement follow too close a cultural-economic line to be held the product of the genes—unless, of course, we are prepared to identify specific gene pools in identified ethnic groups, taking full account of intermarriage, when and how often for various subgroups. As if such things were possible without the convenience of skin color.[28]

In 1921, *The Annals of the American Academy of Political and Social Science* devoted an issue to the problems of child welfare. The issue was a collection of broadly based articles with no reference to socioeconomic factors until, in the concluding section of the volume, an article on the "Problems of the Colored Child" found a place among the "special groups"— the lame, the blind, the orphan, and the illegitimate. The article drew on data from the Blascoer study to make its points, but by now separate attention for blacks indicated a patent awareness of color debilitation on a par with physical and moral crippling. "Negro child study" as a separable entity was symptomatic of the fact that the "usual" agencies found it "impossible to meet the needs" of this racial group, despite the continued use of a language of familial dislocation and disruption closely akin to the language and direction of similar and contemporary studies of city delinquents in general.[29]

Poor school performance was a citywide problem with clearly perceived relationships to social class and economic well-being.

What is intriguing about the separate definition of Negro performance is that in isolating New York's Negro school children,
educators (consciously or unconsciously) were in fact providing
an educational parallel to the life and place of the New York
City Negro on the broader urban stage. Educational marginality
in the public school paralleled marginality in the economy.
Even though private agencies like settlement houses and
churches had to take up the slack in such basic nagging questions as tenement housing and urban health deficiencies, the
faith was that the school could remove the children of the
urban poor from the misery of poverty; and this faith was reinforced because it seemed that the economy could do it.
Black failure and the separate categorization of it revealed the
true measure of the school's limited responsibility for the social
progress of any group. That is why, of course, with the out-
break of World War I and the Red Scare, immigration restriction, not public school mobilization, followed literacy tests.
While concern for the failure of public education for millions
of the foreign-born had culminated in immigration restriction
legislation, little was done to modify the partnership which persisted between lower-class status, school failure, crime, and
delinquency.[30]

Blacks could not be kept out; they came even after the
specific times that they were needed. And so a Harlem district
school superintendent, B. Schlockow, managed to temporarily
forget the Italians in the district, while requesting courses for
teachers that would help them understand the problems con-
fronting Negroes who came to his schools in ever increasing
numbers. But the need to "understand" seemed somehow to
assume unique Negro characteristics, and acted therefore as a
justification for continued black failure in Harlem schools.
These courses were to include a section entitled "the physiology
and psychology of the Negro . . . Special abilities and dis-
abilities—physical, mental and emotional." Whatever the problems presented by the "home environment, . . . economic
status, . . . [and] native ability of children in this district which
hinder the worth of the teachers and of the schools," the big

problem seemed to be that twelve of his schools were already 80 percent black, and that this in itself was a problem to be isolated from the school problems of the urban lower classes.[31]

When Maudelle Bousfield contributed to the first issue of the *Journal of Negro Education* in 1932, she directed her statements, in an issue which was primarily concerned with the South, to officials in Northern urban centers who were then facing "tremendous increase in [Negro] public school enrollment." She urged that the poor preparation of new students be recognized and she recommended that school people develop "special" programs to compensate for it instead of reacting negatively and insensitively as she expected they would. Her fears were, of course, entirely appropriate.[32]

The Mayor's Commission, established in New York City to investigate conditions in Harlem after serious rioting in 1935, poignantly underscored the school's role as an agent of the society and the interminable plight of the Negro. On visitor's day in an elementary school in Harlem it was the custom for the children to play host to parents and touring dignitaries. The tasks of guiding visitors about the school and serving them tea were divided among appointed monitors. The guides were always white girls in nurse's uniforms. Negro girls dressed as waitresses brought the teatime dishes and took them away.[33]

It was found that Negro high school pupils were rarely to be found in the school annexes devoted to college preparatory work. Whether or not the Negro deserved a place in such an institution is almost beside the point, because the fact that other lower-class children were being increasingly prepared for college entrance left more and more Negroes to do the diminishing jobs for which they had been unsuitable twenty years earlier.

What happened during the Depression is crucial to understanding the way in which ethnic identity has been a serious mechanism for misrepresenting the continuity of the school's place in and the character of lower-class urban life, whether sufferers be black or white (always remembering that blacks inherited urban collapse because they are black). After the

1935 Harlem riots, Mayor LaGuardia asked Franklin Frazier to survey the situation and to make recommendations. Desegregation was a basic plank in those recommendations, and the school was its fulcrum. Unlike Forester Washington's 1926 survey following the Detroit riots, or the 1922 Chicago Riot Commission, Frazier's challenge to the schools was more than a simple reiteration of America's faith in public schools. It was an awareness that something new had taken place: at the heart of the Depression, Frazier was responding to the fact that schooling was going to be vital to upward mobility, now as at no other time, and that black exclusion from opportunity in society at large was now well established in those schools too. Something had to change quickly in the schools, or riots might be expected to become the rule.[34]

The Depression was not simply another unfortunate but natural down in the business cycle. It marked the end of unskilled employment as an expanding sector of the economy. The school was no longer to be merely a palliative but an indispensable vehicle for individual and group progress in the city. Stuck in industrial, unskilled labor occupations to which they had only just been admitted, blacks stood no chance of improvement in this new world; segregation in the schools and a dearth of supervisory positions served to typify for Frazier the industrial dead end they had entered. Desegregation was the polemical objective, the political rallying cry, but the goal of its hidden agenda was academic success in school for black pupils in New York and in other urban areas.

Desegregation, as Frazier saw it, meant inclusion of blacks as professionals in a school system which had virtually excluded them since the closing of the "colored schools" at the turn of the century, it meant black professionals in the schools where teachers taught white children their lessons. Frazier believed that once barriers to black opportunity were down, then the school could perform for blacks as it was doing for whites.[35] This view, of course, was the underlying philosophy of the civil rights movement which followed World War II. Taking a general American liberal point of view, blacks accepted the notion that being included was the major problem. Desegregation was a melting-pot notion, a belief in the dissolution of ethnic identity in a new homogeneous American people. Apart from the liberal establishment, which had in effect celebrated the idea of the melting pot since first confronted by large-scale foreign immigration, the blacks were the only group that actively believed in the melting pot.

But desegregation had, as I have suggested, another aspect. And it is this aspect which explains the Negro's loyalty to it for so long. As I have pointed out, emancipation had not meant any change in the status of blacks in the North except the ending of slavery. As slaves they were reserve agrarian labor; in industrial society they served as reserve labor too. Reflecting this status, formal education for blacks took place in "colored schools" in New York City at first, and as these schools were closed, education for blacks was offered in public schools— but as a separate unit of the school population. When school achievement began to be treated seriously as an important prerequisite to job opportunity during the Depression and after World War II, blacks could do little but express their dissatisfaction within the framework of the category established for analyses of problems pertaining to them. In essence, attention was directed at smashing down the category.

More and more blacks came to the city in the migrations following World War I at a time when the long-standing problems of schools were being further complicated by the new demands that the schools take up the slack in the economy—by keeping more students in school for longer periods of time and by accepting a new responsibility for improving academic performance for eventual job opportunity. By the 1930s black children were entering the school in large numbers, and at the same time the inability of the school to make a difference in the academic performance of poor children was becoming a serious social problem. As the school's monopoly on avenues to employment was made increasingly clear after World War II, so its inability to redistribute opportunities fell most heavily on the newest and rapidly the largest part of its student body—the blacks.

The desegregation issue in New York City was soon, as Frazier had anticipated, shorthand for the incidence of unprecedented numbers of black children in school that accompanied the shrinking of the unskilled labor market. In effect, the special category through which the school and social problems of blacks were traditionally examined was expanded to deal with a school system where the majority of poor people's children continued to fail but where the majority of poor people were black. As the campaign around desegregation in schools grew in power in the 1940s and 1950s, so the separate category expanded to a new role—it was now the way we talked about schooling, about equal educational opportunity. Thereby it assumed implicitly that here was a new challenge to the public school: it must now do for blacks what it had done for others. Symbolically that meant for a long time that movement out of black schools, lessons alongside white pupils, was the answer. The slow to negligible progress of poor white children in public schools was conveniently swept under the rug of concern about the Negro. Slow progress among whites in schools and subsequent slow progress in the world of work have, until very recently, obfuscated the association of blacks with the school's inability to "work" for poor children—an association used in earlier decades about much smaller numbers of blacks but serving to explain and justify school failure just the same.[36]

By the mid-1950s the *New York Times Index* changed its "Negro Education" subcategory of its "Education in New York City" section, replacing it with "Racial Integration." The number of pages required to cover it increased so that by the end of the 1950s it occupied as many as all other education categories combined.[37]

The report of the Superintendent of Schools for 1957 showed a drop of 20,000 in the number of white pupils. Negroes then formed a highly concentrated 20.1 percent of the school population. As many as 455 of 704 schools in the city were "homogeneous" with less than 10 percent blacks in white schools and vice versa. In 1954 Kenneth Clark argued that the Brown decision be applied in the schools of the North that were segre-

gated by the effects of a combination of neighborhood concentration of Negroes and neighborhood school zoning. The Board
of Education requested the Public Education Association to
examine the problem. One year later it reported that in schools
of high non-white concentration the children scored below
children of other schools in various achievement tests. The
school buildings in these districts were older and the proportion
of substitute teachers highest.[38]

In 1957 the Board of Education began to keep meticulous
ethnic—"black, white, and other"—records on the racial composition of its schools. By this time blacks constituted 25 percent of elementary school pupils and 20 percent of both junior
high school and vocational high school student bodies. Only 5
percent of the academic high school population was black. Six
years later State Commissioner Allen's study of school segregation found the rate of segregation and poor black performance
increasing and projected the continuation of the trend. Segregation was virtually synonymous with urban school inadequacy.[39]

The tradition of seeing black social and school problems
under separate cover is clearly a form of racism. It serves no
function to overwork the innumerable examples of racism in
the lives of New York blacks. But it is important to remember
that racist attitudes generate a unique method of observation
and analysis. The operation of this method in the public
schools, as among those social school reformers generally concerned with the lives of the poor described at the beginning of
this chapter, is a serious reflection on the limited role of public
schools in improving those lives. The separate black category,
while a function of racism, was also a symptom of the Negro's
place in society and the school's acquiescence to it despite the
good intentions of some individuals. The white poor, too, were
schooled in the same system although for a different place in
it. And the effect of school desegregation on poor whites was
to force them to defend against what seemed to be encroach-
ment on their turf. Meanwhile, the defense against black protest throughout the city and particularly in neighborhoods where
lower-middle-class and middle-level families sought to protect

the gains their hard work had made possible strengthened the school legend enormously, helping to confirm the traditional belief that there was something important to be had in public schools,[40] something that would finally fulfill the promise everyone believed in and so few had experienced.

White parents, too, had been protesting about school conditions in the city since the late 1930s. Here, too, their actual anxieties were hidden behind their expressions of discontent with long-standing conditions—overcrowding, unsafe facilities, long walking distances to school. White parents boycotted schools and established a style of school protest which black parents would follow in desegregation campaigns. But the issue in both white and black protest was really school performance; in both instances the essential efficacy of the schools until that time was taken for granted—the problems were to be solved by external repair work and the final resolution of problems growing out of slavery. The fact was, of course—and the Sputnik scare which occurred in the late 1950s, at almost the same time as the plan to desegregate urban schools, confirmed it—that school performance suddenly loomed large as a factor in opportunity and national economic security. This was a new phenomenon which put unprecedented demands on public schools to facilitate academic achievement among much broader parts of the school population. Wide-scale school failure was—for the first time—no longer functional, and the ethos of public school success supported by its general functionality was brought into serious question. Unfortunately, our naïveté about what school "success" really meant in the past has made it difficult to appreciate the uniqueness of the current demand that public schools make a positive difference in the academic performance of poor children—black or white. Meanwhile, the fact remains that the public education system we have inherited has no precedent whatsoever in its past upon which to draw in order to meet that demand; it has no experience on which to build to do the job we expect it to do.[41]

Conclusion

I have tried in this book to bring a fresh historical perspective to bear on what is to me the embarrassingly simple-minded conventional wisdom about how poorly the public schools are working now. My point is not merely that schools worked poorly in earlier times, but that their failure has been, in fact, a criterion of their social success, then and now. More specifically, it seems to me, the failure of many children has been, and still is, a learning experience precisely appropriate to the place assigned them and their families in the social order. They are being taught to fail and to accept their failure. In the same way, the personal costs paid by those who "succeed" in school (so antithetical to the humane, personal growth rhetoric of the public schools) are a vital part of the learning experience of the more favored pupils. Clearly, it is a mistake to judge the outcome of a socializing agency, the school, according to a set of value judgments which has little social, political, or economic impact in the larger world. If the school's effects are both longstanding and consistent, that must be attributed to the system's success. To evaluate the system by its rhetorical spirit is dangerous and misleading, yet we do it all the time.

The fact of the matter is that American public schools in general, and urban public schools in particular, are a highly successful enterprise. Basic to that success is the high degree of academic failure among students. Attitudes and behavior patterns such as tolerance of boredom, learning as memorization, competition, and hostility are learned and reinforced in the classroom. The schools do the job today that they have always done. They select out individuals for opportunities according to a hierarchical schema which runs closely parallel to existing social class patterns. The problem today is that there is an increasing shortage of even low level employment options for those on the lower levels of the public school totem pole. As a result, the schools now produce people who are a burden upon, rather than the mainstay of, the socio-economic order.

The theory of natural selection which lies behind much of the American popular faith in public education—the notion that given the will and the strength of character, any man can "make it" in American society—this theory consistently ignores the reality and effectiveness of the criteria imposed from above upon those attempting to climb the ladder of success. These criteria, these standards, are clearly designed to impede upward progress of others along those narrow rungs. We know a great deal about how the interests and preferences of elites set the standards for the selection of those below them in the social order, but we choose to downplay what we know in favor of discussing the complexities of cultural and biological inheritance of which we know much less. Since the actual educational power of public education has been so vastly overestimated, perhaps we should really consider the school to be a symbolic mechanism that holds a diverse, highly competitive society together. The school system, then, stands as an institutional statement of public morality, providing a set of defensive guarantees for the protection of the various orders of society. The result, however, is what C. Wright Mills called "liberal rhetoric and the conservative default," which constitute the official liberal line at the heart of the true American dilemma.

My argument in this book is not that a potentially positive public school system has been misused. I am not at all certain about the ability of public schooling to effect radical social change—and really improving the lives of poor people will require radical change. I see the public school to date as a terribly limited, reflexive institution, slavishly serving society rather than leading it. My purpose in challenging the long-standing American faith in public schools, so dear to the hearts of most of our educators and policymakers, is to undermine the two pernicious myths this faith supports. One myth states that public schools did great and marvelous things for poor people in the past. The second myth grows out of the first. It insists that even if the miracles the public schools actually performed in the past were few, they could easily perform one now. Social reformers, social scientists, and schoolmen are convinced of their own good will toward the disadvantaged. When their efforts fail—modest as they may have been—someone else must be at fault. As a result, we blame poor people rather than self-interest, and vested interest, and we use the consequent political stalemates to explain the incapacity of the school to make a positive difference in the lives of those at the bottom of our society.

To make the public schools truly an instrument for creating mobility among poor people requires something quite new. There are no models in the past. Understanding the lack of historical models and the myths surrounding the agency in which we place so much trust for social progress can be a first step in recognizing the truly tentative nature of our commitment to transforming the agonizing inequalities on which the existing social order is built.

Schools could be an agent for major change in this society. Basic as they are to the maintenance of both the humane and democratic rhetoric of society, and of widescale socioeconomic inequalities, the public schools could be a vehicle for some of us to push the contradictions inherent in the severe disjuncture between school rhetoric and the social reality to the point of absurdity. Since the goals in the school's rhetoric are anathema to the organization and theory of a puritan, capitalist society, would it not be absurd and what might happen if that rhetoric became more than words? What if teachers succeeded in promoting literacy and the capacity for critical thinking among those seriously deficient in these skills in a society that hires only a fraction of its work force for jobs requiring such skills? And what if we created in our classrooms a less formal, more humane environment characterized by excitement, respect, and autonomy—by students' participation in the planning and control of learning? Couldn't we expect this more humane environment to reinforce behavioral styles entirely different from the hostile competitiveness presently rewarded and encouraged in the public school classroom? If we could break the harmony that now exists between the schools and society by provoking a change in one of the major ways society fashions itself, might we not be able to contribute to the radical reformation of that society? But for now, as in the past, our schools and society cooperate—all too well.

There is no precedent to comfort the belief that there is hope for society through schools. But for people like myself who have for varied and complex reasons chosen to work in them, maximizing the tension between school rhetoric and school reality seems to be the only viable way to make public education a force for significant social change. Such tension could be the way the public school could contribute to a radical reassessment of social policies despite its continued subservience by and large to economic pressures and powerfully vested interests.

We must demythologize the schools and become much more aware of the tension which exists between the rhetorical goals of schools and this society's clear willingness to encourage massive welfare dependency and poverty rather than do what is necessary to create autonomy and mobility. My firm belief is that in an environment built on an awareness of such tension, many more children will do well academically. Perhaps even more will do well in a framework of mutual respect and acceptance—of competition against self—rather than within the framework of hostility, suspicion, and incompleteness which presently dominates.

If we understand what schools are and try to mobilize them in the direction of what they claim to be—that is, ironically enough, toward the goals they are already popularly identified with—it might be possible to apply real expectations of progress to schools. By actually acting on the implicit and explicit articles of faith (which are now only the pious mouthings of public education officials in this country) we might be able to transform public education, making it truly public—exposing and threatening its institutional, patchwork reform notions of progress which insanely call massive unemployment and underemployment "full employment," and term a 50 percent college dropout rate "universal education."

The important thing to remember is that no matter what they have said about themselves, schools really have no historically sanctified progressive alliance with social change; yet there may be a place for those of us in the schools who consider ourselves radical critics, to stand and fight for what we believe. Our job in schools must be to declare open war on the widespread fear that inclusion of those previously excluded—inclusion of the poor—threatens security and reduces affluence. While this is not the place for a detailed discussion of changes in the national economy, it is the place to call on those of us who say we care about poverty, deprivation, and hostility in America to turn the energy of our concern into a crusade for the classroom equivalent of a full employment economy; a classroom in which the logic of a child's behavior is respected, in which shame, fear, and repressive standardized measures do not dominate. In the absence of such a crusade schools can only be what they have always been. The new criticism of schools in the 1960s, the charge that they are joyless, frigid places which defeat and abort the creative energy of children, is an important one. But we have to be careful to preserve its essential radicalism. And we can only do that if we act on it, not simply add to it. The criticism I refer to, the work of Paul Goodman, John Holt, George Dennison, and Ivan Illich, for example, has been both timely and successful. Following the post-Sputnik panic for hard, scientific curricula, the so-called humane critics of schools joined by dissatisfied blacks and students have forced large urban school systems to listen and include new constituents in their window-dressing campaigns. But we have been here before. The question is whether the plaintiffs or the system will be best served by the integration of the last ten years of criticism into public school rhetoric.

There is a great deal of writing today about schools in the perspective of what has come to be known as a humane, radical critique. My own vision of schools is deeply indebted to it. But I worry about this new literature because all the predecessors of the present human potential movement—the "child centered" movement of the 1920s, the "true democracy" curriculum of the late 1940s and early 1950s—managed to become integrated into classroom practices as techniques designed to continue the successful transmission of those long established conventions having to do with scarcity as the "reality principle" of human intercourse.

The fact is that we have never been short of criticisms of schools. Indeed the patently soft-headed humanitarianism and egalitarianism of the movements just mentioned in the face of the scarcity principle have encouraged many serious critics of public education to blame its periodic difficulties on the presence of too much heart and too little mind. So we have—over the years—those who hold fast and quite insanely to the inevitability of social progress for huddled masses through the schools, and others who declaim against the folly of such a view, so long as reason and intellect continue to take to the sentimental energies which fuel universal education and thus threaten high culture and elites. Together these viewpoints represent a consciousness insured, as R. D. Laing and Herbert Marcuse have put it, to its own falsity. The fact is that heart has been as missing as mind, and the magnificent combination has been missing altogether.

Once again, this is a critical time in education. Today, children are enrolled in school earlier and attend for more years than at any other time or place in history. Demands on education have again reached an unprecedented peak, and have provoked new problems. But the so-called new set of problems is, in fact, a set of old problems, now insistent and uncompromising. They require that we finally take our traditional rhetoric seriously, instead of continuing to look to legends to explain why, in our time, the promise of our society is unfulfilled— and seems destined to remain so.

Notes

INTRODUCTION

1. W. I. Thomas and Florian Znaniecki, *The Polish Peasant in Europe and America,* Vol. 2 (Chicago: University of Chicago Press, 1918; New York: Knopf, 1928; Dover, 1958). Introduction, part IV, p. 1858.

2. See, for example, Barton J. Bernstein, ed., *Towards a New Past: Dissenting Essays in American History* (New York: Pantheon Books, 1968).

3. Samuel Eliot Morison, "Faith of a Historian," *American Historical Review,* 56 (January 1951): 272–273.

4. David B. Tyack, "Forming the National Character: Paradox in Educational Thought of the Revolutionary Generation," *Harvard Educational Review,* 36 (Winter 1966); "Becoming an American: The Education of an Immigrant," in *Turning Points in American Educational History* (Waltham, Mass.: Blaisdell, 1967), Introduction, chapter 7; "The Perils of Pluralism: The Background of the Pierce Case," *American Historical Review,* 74 (October 1968).

5. Maxine Greene, *The Public School and the Private Vision: A Search for America in Education and Literature* (New York: Random House, 1965).

6. Raymond E. Callahan, *Education and the Cult of Efficiency* (Chicago: University of Chicago Press, 1962).

7. Charles E. Strickland and Charles Burgess, eds., *Health, Growth, and Heredity: G. Stanley Hall on Natural Education* (New York: Teachers College Press, Columbia University, 1965).

8. Gilbert Osofsky, *Harlem: The Making of a Ghetto Negro New York, 1890–1930* (New York: Harper and Row, 1965); Seth M.

Scheiner, *Negro Mecca: A History of the Negro in New York City 1865-1920* (New York: New York University Press, 1965).

9. David B. Tyack, "This Period of Ferment May Be a Turning Point," *The New York Times*, January 11, 1971, Annual Education Review.

10. Callahan, *Education and the Cult of Efficiency*.

11. Michael B. Katz, *The Irony of Early School Reform: Educational Innovation in Mid-Nineteenth Century Massachusetts* (Cambridge, Mass.: Harvard University Press, 1968).

12. Gabriel Kolko, *The Triumph of Conservatism* (New York: Free Press, 1963); Roy Lubove, *The Professional Altruist: The Emergence of Social Work as a Career 1880-1930* (Cambridge, Mass.: Harvard University Press, 1965); James Weinstein, *The Corporate Ideal in the Liberal State 1900-1918* (Boston: Beacon Press, 1968).

13. Colin Greer, "Immigrants, Negroes, and the Public Schools," *The Urban Review*, 3 (January 1969), and "Public Schools: Myth of the Melting Pot," *Saturday Review*, 52 (November 15, 1969); David Cohen, "Immigrants in the Schools," *Review of Educational Research*, 40 (February 1970).

CHAPTER 1

1. Henry J. Perkinson, *The Imperfect Panacea: American Faith in Education 1865-1965* (New York: Random, 1968), p. 3.

2. Arthur Lean, "Review of *Public Education of the Future*," *History of Education Journal*, 6 (Fall 1954): 167.

3. Educational Policies Commission, "Public Education and the Future of America" (Washington, D.C., National Education Association and American Association of School Administrators, 1955).

4. Oscar Handlin, *The Uprooted* (New York: Grosset and Dunlap, 1951); Bernard Bailyn, *Education in the Forming of American Society* (New York: Vintage, 1960).

5. Bailyn, *Education in the Forming of American Society*, p. 49.

6. Daniel J. Boorstin, *The Americans: The Colonial Experience* (New York: Vintage, 1958), part 6.

7. "Idea of the English School, Sketch'd Out for the Consideration of the Trustees of the Philadelphia Academy" in Leonard Labarce *et al*, eds., *The Papers of Benjamin Franklin* (New Haven: Yale University Press, 1961), 4, pp. 102-108; Robert Middlekauff, *Ancients and Axioms: Secondary Education in Eighteenth Century New England* (New Haven: Yale University Press, 1963).

8. Paul Leicester Ford, ed., "Notes on Virginia," *Thomas Jefferson, Works*, Vol. 4 (New York, 1904).

9. Rush Welter, *Popular Education and Democratic Thought in America* (New York: Columbia University Press, 1962), pp. 3-6, 335.

10. Ellwood P. Cubberley, *Changing Conceptions of Education* (Boston: Houghton, Mifflin, 1909), p. 68; and *Public Education in the*

United States (Boston: Houghton, Mifflin, 1919), pp. viii, 18–19.

11. Cubberley, *Public Education in the United States*; Lawrence A. Cremin, *The American Common School: An Historic Conception* (New York: Teachers College Press, Columbia University, 1951).

12. Bailyn, *Education in the Forming of American Society*, pp. 47–49; Lawrence A. Cremin, *The Transformation of the School* (New York: Knopf, 1961).

13. Cremin, *Transformation of the School*, Preface, pp. vii–ix, 3–22; and *The Genius of American Education* (New York: Vintage, 1965).

14. Cremin, *Transformation of the School*, Preface, pp. vii–ix, 3–22.

15. Martin S. Dworkin, ed., "Dewey on Education," in *John Dewey: A Centennial Review* (New York: Teachers College Press, Columbia University, 1959), pp. 7–9, 15–18.

16. Nathan Glazer, "Negroes and Jews: The New Challenge to Pluralism," *Commentary*, 38 (December 1964), and "Ethnic Groups in America: From National Culture to Ideology," in H. Berger, T. Abel, and C. H. Page, eds., *Freedom and Control in Modern Society* (New York: Octagon Books, 1964).

17. Perkinson, *Imperfect Panacea*, chapter 3.

18. Cremin, *Genius of American Education*, and *Transformation of the School*; Sol Cohen, *Progressives and Urban School Reform* The Public Education Association of New York City, 1895–1954 (New York: Teachers College Press, Columbia University, 1964); Theodore R. Sizer, *Secondary Schools at the Turn of the Century* (New Haven: Yale University Press, 1964); Daniel J. Boorstin, *The Americans: The National Experience* (New York: Vintage Books, 1965); Edward A. Krug, *The Shaping of the American High School 1880–1920* (Madison, Wisc.: University of Wisconsin Press, 1969).

19. Handlin, *The Uprooted*, p. 222; Gunnar Myrdal, *The American Dilemma* (New York: Harper, 1944); J. Iverne Dowie, "The American Negro: An Old Immigrant on a New Frontier," in O. F. Ander, ed., *In the Trek of· the Immigrant* (Rock Island, Ill.: Augustana College Library, 1964), pp. 241–242; Oscar Handlin, "The Goals of Integration," *Daedalus,* 95 (Winter 1966).

20. Marcus L. Hansen, *The Problem of the Third Generation Immigrant* (Rock Island, Ill.: Augustana Historical Society Publications, 1938).

21. Charles Silberman, "The City and the Negro," *Fortune Magazine,* (March 1962); Miriam L. Goldberg, "Problems in the Evaluation of Compensatory Programs for Disadvantaged Children," *Journal of School Psychology*, 4 (Spring 1966): 113; Dan Dodson, "Education and the Powerless," in A. Harry Passow et al., *Education of the Disadvantaged* (New York: Holt, Rinehart, and Winston, 1967); U.S. Riot Commission, *Report of the National Advisory Commission on Civil Disorders* (New York: Bantam Books, 1968), pp. 279–281.

22. Fred Hechinger, "School Vouchers: Can the Plan Work?" *The New York Times*, June 7, 1970; Albert Shanker, "The Big Lie About the Public Schools," *The New York Times*, May 9, 1971, U.F.T. Column.

23. Harold Cruse, *The Crisis of the Negro Intellectual* (New York: Morrow, 1967); Eldridge Cleaver, *Soul on Ice* (New York: Ramparts/McGraw-Hill, 1968); Ernest Fergusson, *Baltimore Sun*, May 6, 1969; Kenneth B. Clark, *Dark Ghetto* (New York: Harper and Row, 1965), and ed., *Racism and American Education* (New York: Harper/Colophon, 1970); Thomas Fellows, "Educating Young Blacks," *The New York Times*, May 14, 1971, Letter to the Editor.

24. Daniel P. Moynihan, et al., "The Schools in the City," *Harvard Today* (Autumn, 1967): 19.

25. Edward Banfield, *The Unheavenly City* (Boston: Little, Brown, 1970).

26. Paul Goodman, *Compulsory Mis-education* (New York: Vintage, 1964), pp. 179–180.

27. Oscar and Mary Handlin, "Mobility," in Edward N. Saveth, ed., *American History and the Social Sciences* (New York: Free Press, 1964), pp. 222–223.

28. Henry Steele Commager, *The American Mind* (New Haven: Yale University Press, 1950), quoted in Peter Schrag, *Voices in the Classroom* Public Schools and Public Attitudes (Boston: Beacon, 1965), pp. 1–2.

CHAPTER 2

1. Iverne Dowie, "The American Negro: An Old Immigrant on a New Frontier," in O. F. Ander ed., *In the Trek of the Immigrant* (Rock Island, Ill.: Augustana College Library, 1964); Oscar Handlin, "The Goals of Integration," *Daedalus*, 95 (Winter 1966).

2. Timothy L. Smith, "Native Blacks and Foreign Whites: Varying Responses to Educational Opportunity in America 1880–1950," paper presented at the Annual Meeting of the American Education Research Association, Spring 1970 (Mimeographed).

3. E. Franklin Frazier, *The Negro in the United States* (New York: Macmillan, 1957), chapters 12 and 13; Charles Silberman, *Crisis in Black and White* (New York: Random House, 1964); Handlin, "The Goals of Integration"; Robert J. Havighurst and Berenice L. Neugarten, *Society and Education* (Boston: Allyn and Bacon, 1967).

4. For the existing opposite picture, see: James Coleman *et al, Equality of Educational Opportunity* (Washington, D.C.: Government Printing Office, 1966); Robert Rosenthal and Lenore Jacobson, *Pygmalion in the Classroom* (New York: Holt, Rinehart and Winston, 1968); Arthur R. Jensen, "How Much Can We Boost IQ and Scholastic Achievement," *Harvard Educational Review*, 39 (Winter 1969).

5. Martin Mayer, *The Schools* (New York: Harper and Row, 1961); Bert Swanson, *Struggle for Equality: School Integration Controversy in New York City* (New York: Hobbs, Dormen, 1966); Marilyn Gittell, *Participants and Participation in New York City* (New York: Center for Urban Education, 1967); David Rogers, *110 Livingston Street* (New York: Random House, 1968).

6. Patricia Sexton, *Education and Income: Inequalities in Our Public Schools* (New York: Viking Press, 1961); Colin Greer, "A View from Coney Island," *The Center Forum*, 2 (December 20, 1967); U.S. Department of Labor, "A Sharper Look at Unemployment in U.S. Cities and Slums" (January 1967); U.S. Department of Labor, Urban Employment Survey, Report No. 1, "Poverty—The Broad Outline, Detroit" (February 1970); U.S. Department of Labor, Urban Employment Survey, Report No. 2, "Poverty—The Broad Outline, Chicago" (March 1970); U.S. Department of Labor, "The New York Puerto Rican: Patterns of Work Experience," Bureau of Labor Statistics, Regional Report No. 19 (May 1971); Lawrence A. Mayer, "Young America: By the Numbers" in "The Stairway of Education," *Fortune Magazine*, 77 (January 1968); S. M. Miller and Pamela Roby, *The Future of Inequality* (New York: Basic Books, 1970); Martin Rein, *Social Policy* (New York: Random House, 1970), Parts 1, 4; National Urban Coalition, *Counterbudget: A Blueprint for Changing National Priorities* (New York: Praeger, 1971).

7. Greer, "A View from Coney Island"; author's personal communication with Dolores G. Chitraro and Max Myers, both assistant superintendents for school districts in Brooklyn, New York, 1967.

8. Miller and Roby, *Future of Inequality*; Sexton, *Education and Income*; Peter Schrag, *Voices in the Classrooms: Public Schools and Public Attitudes* (Boston: Beacon Press, 1965), p. 1; Theodore R. Sizer, "The Schools in the City," *Harvard Today* (Autumn 1967); Stephen Thernstrom, "Poverty in Historical Perspectives," in Daniel P. Moynihan, ed., *On Understanding Poverty* (New York: Basic Books, 1969).

9. U.S. Bureau of the Census, Current Population Reports, Series P-20, No. 207, "Educational Attainment: March 1970" (1971).

10. Miller and Roby, *Future of Inequality*; Daniel Schreiber, ed., *Profile of the School Dropout* (New York: Vintage, 1968); A. J. Jaffe and Walter Adams, "American Higher Education in Transition" (New York: Bureau of Applied Social Research, Columbia University, 1969); Ivar Berg, *Education and Jobs: The Great Training Robbery* (New York: Praeger, 1970).

11. Havighurst and Neugarten, *Society and Education*, chapter 12; David N. Alloway and Francesco Cordasco, *Minorities and the American City* (New York: Vintage, 1968).

12. *Ibid.*

13. Coleman, *Equality of Educational Opportunity*; Sexton, *Education & Income*.

14. See *The New York Times* Annual Education Review, January 11, 1971.

15. Coleman, *Equality of Educational Opportunity*; Jaffe and Adams, "American Higher Education in Transition"; Algo G. Henderson, *Policies and Practices in Higher Education* (New York: Harper and Brothers, 1960), chapter 4; John Sommerskill, "Dropouts from College," in Nevitt Sanford, *The American College* (New York: Vintage, 1962); John Holt, *How Children Fail* (New York: Pitmans, 1965); James Trent and Leland Medsker, *Beyond High School* (San Francisco: Jossey-Bass, 1966); Arthur M. Cohen, *Dateline '79: Heretical Concepts for the Community College* (Beverly Hills, Calif., 1969); U.S. Bureau of the Census, Current Population Reports, "Educational Attainment: March, 1970" (1971); Pete Hamill, "The White High Schools," *New York Post* (March 11, 1970); Robert Schrank and Susan Skein, "Yearning, Learning, and Status," in Sar Levitan, ed., *Blue Collar Blues* (New York: McGraw-Hill, 1971).

16. Gabriel Kolko, *Wealth and Power in America: An Analysis of Social Class and Income Distribution* (New York: Praeger, 1962), chapter 4; Peter Blau and Otis D. Duncan, *The American Occupational Structure* (New York: Wiley, 1967), chapter 5; Christopher Jencks and David Riesman, *The Academic Revolution* (New York: Doubleday, 1968), p. 79; Bureau of the Census, Current Population Reports, Series P-20, No. 185, July 11, 1969; Seymour Lipset and Reinhold Bendix, *Social Mobility in Industrial Society* (Berkeley, Calif.: University of California Press, 1969).

17. Fred Hechinger, "CUNY Begins Vital Test of Open Admissions," *The New York Times*, September 20, 1970; Robert Birnbaum and Joseph Goldman, "The Graduates: A Follow-Up Study of New York City High School Graduates of 1970" (New York: City University of New York, 1971).

18. Greer, "A View from Coney Island"; Marcus Hansen, *The Problem of the Third Generation Immigrant* (Rock Island, Ill.: Augustana Historical Society Publications, 1938), pp. 1–4; Leonard Covello and Guido D'Agostino, *The Heart Is the Teacher* (New York: McGraw-Hill, 1958); Nathan Glazer and Daniel B. Moynihan, *Beyond the Melting Pot: The Negroes, Puerto Ricans, Jews, Italians, and Irish of New York City* (Cambridge, Mass.: M.I.T. Press, 1963); H. McNeill *et al,* "Demographic Information by Health Area" (New York: Maimonides Medical Center, Program Evaluation Section, September, 1967; mimeographed); U.S. Bureau of the Census, Current Population Reports, Series P-20, No. 220, "Ethnic Origin and Educational Attainment: November, 1969"; Peter Schrag, "Is Main Street Still There?" *Saturday Review,* 53 (January 17, 1970); New York City Commission on Human Rights, "Equal Employment Opportunity and the New York City Schools" (January 25–29, 1971); Peter Schrag, "Growing Up on Mechanic Street," in *Out of Place in America* (New York: Random House, 1971).

CHAPTER 3

1. Robert L. Heilbroner, *The Future as History* (New York: Harper and Row, 1959), pp. 30–35.

2. Louis Hartz, *The Liberal Tradition in America* (New York: Harcourt, Brace, 1955); Rush Welter, *Popular Education and Democratic Thought in America* (New York: Columbia University Press, 1962), Conclusion; Allen Hansen, *Liberalism and American Education in the Eighteenth Century* (New York: Octagon, 1965).

3. Edward Shils, "Plenitude and Scarcity: The Anatomy of an International Cultural Crisis," *Encounter* (May 1969): 37–40.

4. Hartz, *Liberal Tradition in America*; Louis Hartz, *The Founding of New Societies* (New York: Harbinger Books, 1964).

5. Philip H. Phenix, *Realms of Meaning* (New York: McGraw-Hill, 1964).

6. Ellwood P. Cubberley, *Public Education in the United States* (Boston: Houghton, Mifflin, 1919).

7. Bernard Bailyn, *Education in the Forming of American Society* (New York: Vintage, 1960); Lawrence A. Cremin, *The Genius of American Education* (New York: Vintage, 1965).

8. Cubberley, *Public Education in the United States*; Bailyn, *Education in the Forming of American Society*; Raymond Callahan, *Education and the Cult of Efficiency* (Chicago: University of Chicago Press, 1962).

9. Merle Curti, *The Social Ideas of American Educators* (Toronto: Littlefield, Adams, 1961).

10. Bailyn, *Education in the Forming of American Society*; Cremin, *Genius of American Education*.

11. Richard J. Storr, *The Role of Education in American History* (New York: Fund for the Advancement of Education, 1957); William W. Brickman, "Revisionism and the Study of the History of Education," *History of Education Quarterly* 4 (1964); Committee on the Role of Education in American History *Education and American History* (New York: Fund for the Advancement of Education, 1965).

12. Bailyn, *Education in the Forming of American Society*, pp. 8–9.

13. Cremin, *Genius of American Education*; Lawrence A. Cremin, *The Wonderful World of Ellwood Patterson Cubberley* (New York: Teachers College Press, Columbia University, 1965).

14. Lawrence A. Cremin, *The Transformation of the School* (New York: Knopf, 1961); Michael B. Katz, *The Irony of Early School Reform: Educational Innovation in Mid-Nineteenth Century Massachusetts* (Cambridge, Mass.: Harvard University Press, 1968), Preface.

15. Bernard Bailyn, "Some Historical Notes," in John Watton and James L. Kuethe, eds., *The Discipline of Education* (Madison, Wisc.: University of Wisconsin Press, 1963), pp. 41–43.

16. Cremin has frequently remarked about the oversimplified picture portrayed in this early work in seminars and classes at Teachers College, Columbia University.
17. Bailyn, *Education in the Forming of American Society.*
18. *Ibid.*
19. *Ibid.*, p. 39.
20. Bailyn, *Education in the Forming of American Society*, pp. 30–31, 43–45.
21. For information on the current power of the family in terms of academic employment opportunity and performance, see: James Coleman *et al, Equality of Educational Opportunity* (Washington, D.C.: Government Printing Office, 1966); John Coons *et al, Private Wealth and Public Education* (Cambridge, Mass.: Harvard University Press, 1970); S. M. Miller and Pamela Roby, *The Future of Inequality* (New York: Basic Books, 1970).
22. Bailyn, *Education in the Forming of American Society.*
23. Cremin, *Genius of American Education*, Preface.
24. *Ibid.*, pp. 4–7.
25. *Ibid.*, pp. 7–9.
26. *Ibid.*, p. 6.
27. *Ibid.*
28. *Ibid.*, p. 5.
29. Cremin, *Transformation of the School.*
30. Clarence Karier, "Review of Patricia Albjerg Graham's *Progressive Education: From Arcady to Academe*," *Educational Theory*, 20 (Spring 1970); Michael Zuckerman, "Review of Lawrence Cremin's *American Education: The Colonial Experience, 1607–1787*," *American Association of University Professors Bulletin*, 57 (Spring 1971).
31. Bailyn, *Education in the Forming of American Society*; Cremin, *Genius of American Education*, and *Transformation of the School.*
32. *Ibid.*
33. Lawrence A. Cremin, "Review of Bernard Bailyn's *Education in the Forming of American Society*," *History of Education Quarterly*, 4 (1964).
34. Ivan Illich, *Deschooling Society* (New York: Harper and Row, 1971).
35. Henry Steele Commager, *The American Mind* (New Haven: Yale University Press, 1950).

CHAPTER 4

1. Louis Hartz, *The Liberal Tradition in America* (New York: Harcourt, Brace, 1955), and *The Founding of New Societies* (New York: Harbinger Books, 1964), Introduction.
2. Gabriel Kolko, *The Triumph of Conservatism* (New York: Free Press, 1963); James Weinstein, *The Corporate Ideal in the Liberal*

State 1900–1918 (Boston: Beacon, 1968); William Appleman Williams, *The Tragedy of American Diplomacy* (New York: Dell, 1970).

3. Erik H. Erikson, *Young Man Luther* (New York: Norton, 1958), pp. 34–35.

4. Maxine Greene, *The Public School and the Private Vision: A Search for America in Education and Literature* (New York: Random House, 1965), pp. 31–37.

5. Greene, *Public School and the Private Vision*; Lawrence A. Cremin, ed., *The Republic and the School: Horace Mann on the Education of Free Men* (New York: Teachers College Press, Columbia University, 1957).

6. Cremin, *Republic and the School*; Greene, *Public School and the Private Vision*, pp. 46–58; Merle Curti, *The Social Ideas of American Educators* (Toronto: Littlefield, Adams, 1961).

7. Robert Ernst, *Immigrant Life in New York City, 1825–1863* (New York: I. J. Friedman, 1965); Roy Lubove, *The Professional Altruist: The Emergence of Social Work as a Career 1880–1930* (Cambridge, Mass.: Harvard University Press, 1965); Allan Horlick, "Counting Houses and Clerks: A Study of the Social Control of Young Men in New York 1840–1950" (Unpublished Doctoral Dissertation, University of Wisconsin, 1969); David J. Rothman, *The Discovery of the Asylum: Social Order and Disorder in the New Republic* (Boston: Little, Brown, 1971), pp. 109–120, 155–179, 206–264.

8. Ernst, *Immigrant Life in New York City*.

9. Curti, *Social Ideas of American Educators*.

10. Louis Hartz, "The Reactionary Enlightenment," *Western Political Quarterly*, 5 (March, 1952); Barrington Moore, *Social Origins of Dictatorship and Democracy; Lord and Peasant in the Making of the Modern World* (Boston: Beacon, 1966), chapter on the United States; C. Vann Woodward, "White Racism and Black 'Emancipation,'" *New York Review of Books* (February 27, 1969).

11. Alexis de Tocqueville, *Democracy in America* (New York: Vintage, 1959).

12. Eugene H. Berwanger, *The Frontier Against Slavery: Western Anti-Negro Prejudice and the Slavery Extension Controversy* (Urbana, Ill.: University of Illinois Press, 1969).

13. Curti, *Social Ideas of American Educators*; Woodward, "White Racism and Black 'Emancipation'"; C. Vann Woodward, *The Burden of Southern History* (New York: Vintage, 1960).

14. Woodward, "White Racism and Black 'Emancipation'"; Eileen S. Kraditor, *Means and Ends in American Abolitionism: Garrison and His Critics, 1834–1850* (New York: Pantheon, 1969).

15. Woodward, "White Racism and Black 'Emancipation'"; V. J. Voegeli, *Free But Not Equal: The Midwest and the Negro During the Civil War* (Chicago: University of Chicago Press, 1969), pp. 57–65.

16. Frederick Jackson Turner, *The Frontier in American History* (New York: Holt, Rinehart, and Winston, 1962).

17. Kraditor, *Means and Ends in American Abolitionism*; Colin Greer, *Cobweb Attitudes: Essays on American Education and Culture* (New York: Teachers College Press, Columbia University, 1970).

18. Lawrence A. Cremin, *The American Common School: An Historic Conception* (New York: Teachers College Press, Columbia University, 1951); Arthur Schlesinger, Jr., *Political and Social History of the U.S., 1829–1924* (New York: Macmillan, 1951).

19. Marvin Meyers, *The Jacksonian Persuasion: Politics and Belief* (Stanford, Calif.: Stanford University Press, 1957); Lee Benson, *The Concept of Jacksonian Democracy* (Princeton, N.J.: Princeton University Press, 1961).

20. Michael B. Katz, *The Irony of Early School Reform* (Cambridge, Mass.: Harvard University Press, 1968), pp. 27–49.

21. *Ibid.*

22. Curti, *Social Ideas of American Educators*; Greene, *Public School and the Private Vision*.

23. Curti, *Social Ideas of American Educators*, pp. 75–77; Greene, *Public School and the Private Vision*, pp. 146–154.

24. Greene, *Public School and the Private Vision*, pp. 31–37.

25. Michael B. Katz, "From Voluntarism to Bureaucracy in U.S. Education" (1970; mimeographed); Clarence Karier, "Elite Views on American Education," in Walter LaQueur and George L. Mosse, eds., *Journal of Contemporary Education*, 6 (New York: Harper Torchbooks, 1970); Clarence Karier, "Testing for Order and Control in the Corporate State" (Urbana, Ill.: University of Illinois Press, 1971; mimeographed).

26. Horace Mann, "Seventh Annual Report of the Board of Education Together with the Seventh Annual Report of the Secretary of the Board" (Boston, 1844).

27. Greene, *Public School and the Private Vision*; Richard Hofstadter, *Anti-Intellectualism in American Life* (New York: Knopf, 1966).

28. Daniel Webster as quoted in Charles Burgess, "The Educational State in America" (Unpublished Doctoral Dissertation, 1962). Also: Daniel Webster, *Works*, Vol. 1 (Boston: Little, Brown, 1954); Curti, *Social Ideas of American Educators*, pp. 46–58; Greene, *Public School and the Private Vision*.

29. For a middle-class rather than factory-worker core of urban rebellion in Europe, see: William Lanzer, "The Pattern of Urban Revolution in 1848," in Evelyn L. Acomb and Marvin L. Brown, Jr., eds., *French Society and Culture Since the Old Regime* (New York: Holt, Rinehart and Winston, 1966); George Rude, *The Crowd in the French Revolution* (Oxford: Clarendon Press, 1967); William Lanzer, *Political and Social Upheaval, 1832–1852* (New York: Harper and Row, 1969). See also: Merle Curti, "The Impact of the Revolutions of 1848 on American Thought," *Proceedings of the American Philosophical Society*, 93 (June 1949).

30. For a brilliant analysis of the "peaceful" transition to industrialism and middle-class culture in Britain, to which I am greatly indebted, see: Perry Anderson, "Components of the National Culture," in A. Cockburn and R. Blackburn, eds., *Student Power* (Hammondsworth, England: Penguin Books, 1969).

31. Lubove, *Professional Altruist*; H. G. Wells, *The New Machiavellians* (Garden City, N.Y.: Doubleday, Duncan, 1932), particularly his "Portrait of Beatrice Webb"; Kitty Muggeridge, *Beatrice Webb: A Life, 1858–1943* (London: Secker and Warburg, 1967); Martin Rein, *Social Policy* (New York: Random House, 1970), chapter 15.

32. Lubove, *Professional Altruist*; Paul Violas, "Jane Addams and Social Control" (Urbana, Ill.: University of Illinois, 1969; mimeographed); Clarence J. Karier, "Business Values and the Educational State" (Urbana, Ill.: University of Illinois, 1971; mimeographed), pp. 18–25.

33. Clarence Karier, "Business Values and the Educational State," pp. 21–23, and "Elite Views on American Education"; Jane Addams, "The Public School and the Immigrant Child," in National Education Association, *Journal of Proceedings and Addresses* (1908).

34. W. A. C. Stewart, *The Educational Innovators: Progressive Schools, 1881–1967* (London: Macmillan, 1968).

CHAPTER 5

1. Raymond Callahan, *Education and the Cult of Efficiency* (Chicago: University of Chicago, 1962); Seth M. Scheiner, *Negro Mecca: A History of the Negro in New York, 1865–1920* (New York: New York University Press, 1965); Neil G. McCluskey, ed., *Catholic Education in America: A Documentary History* (New York: Teachers College Press, 1966); David B. Tyack, "Religious Folkways and the Law of the Land," in Paul Nash, ed., *History and Education* (New York: Random House, 1970).

2. Ellwood P. Cubberley, *Public Education in the United States* (Boston: Houghton, Mifflin, 1919), pp. 211–212.

3. Callahan, *Education and the Cult of Efficiency*; Sol Cohen, *Progressives and Urban School Reform: The Public Education Association of New York City, 1895–1954* (New York: Teachers College Press, Columbia University, 1964); David B. Tyack, "City Schools at the Turn of the Century: Centralization and Social Control" (Stanford, Calif.: School of Education, Stanford University, 1969; mimeographed); Peter Schrag, "The Decline of the WASP," *Harper's Magazine* (April 1970); "The Foreign Element in New York City," *Harper's Magazine* (1890).

4. Cohen, *Progressives and Urban School Reform*; Robert E. Park, *The Immigrant Press and Its Control* (New York: Harper, 1922); Theodore Lowi, *At the Pleasure of the Mayor: Patronage and Power in New York City, 1898–1958* (New York: Free Press,

1964), p. 72; Roy Lubove, *The Professional Altruist: The Emergence of Social Work as a Career, 1880–1930* (Cambridge, Mass.: Harvard University Press, 1965); James Weinstein, *The Corporate Ideal in the Liberal State, 1900–1918* (Boston: Beacon Press, 1968).

5. Oscar Handlin, "The Goals of Integration," *Daedalus*, 95 (Winter 1966); William Ryan, *Blaming the Victim* (New York: Pantheon, 1971).

6. Nathan Glazer and Daniel P. Moynihan, *Beyond the Melting Pot: The Negroes, Puerto Ricans, Jews, Italians, and Irish of New York City* (Cambridge, Mass.: M.I.T. Press, 1963), pp. 50–53.

7. Moses Rischin, *The Promised City: New York's Jews, 1870–1914* (Cambridge, Mass.: Harvard University Press, 1962); H. McNeill *et al*, "Demographic Information by Health Area" (New York: Maimonides Medical Center, Program Evaluation Section, 1967; mimeographed).

8. Lloyd W. Warner and Leo Srole, *The Social Systems of American Ethnic Groups* (New Haven: Yale University Press, 1964); Andrew Greeley, *Why Can't They Be Like Us? America's White Ethnic Groups* (New York: Dutton, 1971).

9. Walter Laidlaw, *Statistical Sources for Demographic Studies, Greater New York, 1910* (New York: New York Federation of Churches, 1912), p. 183; and *1920* (New York Federation of Churches, 1922), pp. 45, 52; Niles Carpenter, "Immigrants and Their Children," Census Monograph No 7 (Washington, D.C.: Government Printing Office, 1927), Vol. 1, Table 16, p. 32; Masakazo Iwata, "The Japanese Immigrants in California Agriculture," *Agricultural History* (January 1962); Theodore Saloutos, "Exodus—U.S.A.," in O. F. Ander ed., *In the Trek of the Immigrant* (Rock Island, Ill.: Augustana College Library, 1964), pp. 199–201.

10. Oscar Handlin, *The Uprooted* (New York: Grosset and Dunlap, 1951).

11. W. I. Thomas and Florian Znaniecki, *The Polish Peasant in Europe and America* (Chicago: University of Chicago Press, 1918; New York: Knopf, 1928, Dover, 1958); Tora Bøhn, "A Quest for Norwegian Folk Art in America," *Norwegian-American Studies and Records,* 19 (1956); Rose Hum Lee, *The Chinese in the United States of America* (Cambridge: Oxford University Press, 1960); Alex Simirenko, *Pilgrims, Colonists and Frontiersmen: An Ethnic Community in Transition* (New York: Free Press, 1964); Rudolph J. Vecoli, "Contadini in Chicago: A Critique of *The Uprooted*," *Journal of American History* (December 1965).

12. Rischin, *Promised City*.

13. Isaac A. Hourwich, *Immigration and Labor* (New York: B. W. Huebsch, Inc., 1922).

14. Jane Addams, *Forty Years at Hull House* (New York: Macmillan, 1929), Introduction by Lillian Wald; Lillian Wald, *The House on Henry Street* (New York: Holt, 1915); Robert E. Park and Herbert A. Miller, *Old World Traits Transplanted* (New York: Harper,

1921); Henry Pratt Fairchild, ed., *Immigrant Backgrounds* (New York: Wiley, 1927); Jacob Riis, *How the Other Half Lives: Studies Among the Tenements of New York* (New York, 1890; Sagamore Press, 1957).

For continual accounts of the conditions of immigrant life and its causes, see: *Charities* (1899 through 1909); *Crisis* (1910 through 1921); *Milestones* (1893 through 1898, 1903 through 1910); *Outlook* (1906, 1914); *Survey* (1909 through 1920).

15. Carpenter, "Immigrants and Their Children"; Glazer and Moynihan, *Beyond the Melting Pot*, Tables; Lloyd W. Warner, Robert J. Havighurst, and Martin B. Loeb, *Who Shall Be Educated?* (New York: Harper, 1944); Christopher Jencks and David Riesman, "On Class in America," *Public Interest*, 10 (Winter 1968).

16. John J. Kane, "The Social Structure of American Catholics," *The American Catholic Sociological Review*, 16 (March 1955): 30; Bosco D. Cestello, "Catholics in American Commerce and Industry, 1925–1945," *The American Catholic Sociological Review*, 17 (October 1956); E. P. Hutchinson, *Immigrants and Their Children, 1850–1950* (New York: Wiley, 1956); Gerhard Lenski, *The Religious Factor* (Garden City, N.Y.: Doubleday, 1961); Stanley Lieberson, *Ethnic Patterns in American Cities* (New York: Free Press, 1963).

17. Kate Haliday Claghorn, "Immigration and Its Relation to Pauperism," *The Annals of the American Academy of Political ana Social Science*, 24 (October 1904); U.S. Immigration Commission "Reports," Senate Document No. 747 (1911), p. 620; Kate Haliday Claghorn, "First Year's Work of a New State Bureau," *The Survey*, 15 (May 11, 1912); U.S. Department of Labor, "Annual Report of Commissioner General of Immigration" (Washington, D.C., 1921), p. 190; David Cohen, "Immigrants in the Schools," *Review of Educational Research*, 40 (February 1970); Hourwich, *Immigration and Labor*.

18. Paul Hanus, "Survey of New York City Schools, 1911" (New York City Board of Education, 1911); George D. Strayer, "Age and Census of Schools and Colleges: A Study of Retardation and Elimination," U.S. Bureau of Education Bulletin No. 5 (1911); U.S. Bureau of Education, "Report of the Committee of the National Council of Education: Standards and Tests for Measuring the Efficiency of Schools or Systems of Schools," Bulletin No. 13 (1913), pp. 3–5; George D. Strayer, "The Paterson School Survey" (New York: Teachers College, Columbia University, 1918), pp. 21–25; William H. Allen, "School Survey Committee Report—The Allen Report" (New York City Board of Education, 1925), pp. 281, 1179–1191; George D. Strayer, "Foreword to the Fundamentals in Education: The Hartford School" (New York: Teachers College, Columbia University, 1937); Basil Coleman, *The Coleman Report: Statistics on Public Schools, 1937–1938* (Washington, D.C.: Government Printing Office, 1938); pp. 13–18.

19. Glazer and Moynihan, *Beyond the Melting Pot*, Tables, pp. 222–224; Carpenter, "Immigrants and Their Children," Vol. 1, Table 16, p. 32; Laidlaw, *Statistical Sources for Demographic Studies, Greater New York, 1910* and *1920*, pp. 52–76; Jencks and Riesman, "On Class in America," p. 79; Walter Laidlaw, "Population of the City of New York, 1890–1930 (New York Census Committee, 1932), p. 185; Bureau of the Census, Current Population Reports, Series P-20, No. 220, "Ethnic Origin and Educational Attainment, November, 1969" (1970), 1969), pp. 7–8. Robert Coles, *Still Hungry in America* (New York: New American Library, 1969); Colman McCarthy, "40 Million Americans and a Broken Odyssey," *Washington Post*, July 13, 1970; *New York Times*, July 13, 1970, September 20, 1970, October 5, 1970, October 8, 1970.

20. *See*: S. M. Miller and Pamela Roby, *The Future of Inequality* (New York: Basic Books, 1970), p. 133; Glazer and Moynihan, *Beyond the Melting Pot*, p. 322, 324; U.S. Bureau of the Census, Current Population Reports, Series P-20, No. 207, "Educational Attainment, March 1970" (1971), p. 7. See also: H. McNeill *et al*, "Demographic Information by Health Area"; *The New York Times*, September 20, 1970.

In New York City, the center of the "new" immigration and melting pot imagery, more than half of the new students in the City University Open Enrollment Program, through which all high school graduates would have access to college, were from white working-class groups, the first members of their families—very often—to go to college. The Maimonides Survey showed that greatly underestimated numbers of Jews and Catholics still occupy low level employment and miserable living conditions.

21. Greeley, *Why Can't They Be Like Us?*; Horace Kallen, "Democracy Versus the Melting Pot," *Nation* (1915); Isaac B. Berkson, *Theories of Americanization* (New York: Teachers College, Columbia University, 1920); Milton M. Gordon, *Assimilation in American Life* (London: Oxford University Press, 1964).

22. Gordon, *Assimilation in American Life*.

23. *Ibid.*, pp. 105–117; Will Herberg, *Catholic-Protestant-Jew* (New York: Doubleday, 1955).

24. Greeley, *Why Can't They Be Like Us?*; Lieberson, *Ethnic Patterns in American Cities*; Nathan Glazer, "Ethnic Groups in America: From National Culture to Ideology," in M. Berger, T. Abel, and C. H. Page, eds., *Freedom and Control in Modern Society* (New York: Octagon Books, 1964).

25. Ruth Miller Elson, "American Schoolbooks and Culture in the Nineteenth Century," *Mississippi Valley Historical Review*, 46 (December 1959), and "Immigrants and Schoolbooks in the Nineteenth Century" (History of Education Society Eastern Regional Meeting, April 1971); Colin Greer, "Attitudes Toward the Negro in New York City, 1890–1914" (Unpublished Master's Thesis, London University, 1968).

26. Greer, "Attitudes Toward the Negro in New York City"; Lubove, *Professional Altruist*; Paul Violas, "Jane Adams and Social Control" (Urbana, Ill.: University of Illinois, 1969; mimeographed). *The New Republic* (January 29, 1916) editorial typifies the Progressive liberal's selective observation of the contemporary social structure and his precarious balance between immigration and restriction, until World War I provoked fear of internal disunity and those still "new" to America became a "special" instance of breakdown in the hitherto effective assimilation apparatus: "Only recently . . . we had absolute confidence in our power of assimilation. Serb, American, Lithuanian, we assured ourselves, would put off their national characters and become good Americans . . . as . . . Irish, Germans and Scandinavians had become merged with the original English stock . . . This optimism is hard to remember."

Peter Schrag, *The Decline of the WASP* (New York: Random House, 1971).

27. Leonard Covello and G. D'Agostino, *The Heart Is the Teacher* New York: McGraw-Hill, 1958); Colin Greer, *Cobweb Attitudes: Essays on American Education and Culture* (New York: Teachers College Press, Columbia University, 1970).

28. Greeley, *Why Can't They Be Like Us?*, p. 47.

29. Blau and Duncan, *American Occupational Structure*; Glazer and Moynihan, *Beyond the Melting Pot*, Tables. Robert Coles, *Still Hungry in America* (New York: New American Library, 1969); Colman McCarthy, "40 Million Americans and a Broken Odyssey," *Washington Post*, July 13, 1970; *New York Times*, July 13, 1970, September 29, 1970, October 5, 1970, October 8, 1970.

30. Bøhn, "Quest for Norwegian Folk Art in America"; Lee, *Chinese in the United States of America*; Glazer and Moynihan, *Beyond the Melting Pot*; Rischin, *Promised City*; Simirenko, *Pilgrims, Colonists, and Frontiersmen*; Thomas and Znaniecki, *Polish Peasant in Europe and America*; Vecoli, "Contadini in Chicago"; Theodore Saloutos, *The Greeks in the United States* (New York: Teachers College Press, Columbia University, 1967).

31. William Shannon, *The American Irish* (New York: Collier-Macmillan, 1963), p. 19.

32. Karen Larson, "Review of Oscar Handlin's *The Uprooted*," *American Historical Review* (April 1952); Vecoli, "Contadini in Chicago."

33. Vecoli, "Contadini in Chicago," pp. 404–417.

34. Alan Conway, *The Welsh in America: Letters from the Immigrants* (Minneapolis: University of Minnesota Press, 1961); Wayne G. Broehl, Jr., *The Molly Maguires* (Cambridge, Mass.: Harvard University Press, 1964).

35. Glazer, "Ethnic Groups in America"; A. William Haglund, "Finnish Immigrant Farmers in New York, 1910–1960," in Ander, ed., *In the Trek of the Immigrant* (Rock Island, Ill.: Augustana Col-

lege Library, 1964); John A. Hawgood, *The Tragedy of German-America* (New York: Putnam, 1940).

36. Rischin, *Promised City*, pp. 51–75; Rudolph Glanz, *Jew and Irish: Historic Group Relations and Immigration* (New York: 1966).

37. Saloutos; *Greeks in the United States*, pp. 33–61, 199–201.

38. Greer, *Cobweb Attitudes*; Howard Brotz, *The Black Jews* (New York: Free Press, 1964); Herbert Aptheker, ed., "Some Unpublished Writings of W. E. B. DuBois," *Freedomways*, 5 (Winter 1965): 103–110, 117–118.

39. Handlin, "The Goals of Integration," and *The Uprooted*; Edmund D. Cronon, *Black Moses: The Story of Marcus Garvey and the Universal Negro Improvement Association* (Madison: University of Wisconsin Press, 1955).

40. Blau and Duncan, *American Occupational Structure*; Marc Fried, "Deprivation and Migration: Dilemmas of Causal Interpretation," in Daniel P. Moynihan, ed., *On Understanding Poverty* (New York: Basic Books, 1969).

41. Fried, "Deprivation and Migration."

42. Handlin, "The Goals of Integration," pp. 268–284; Edward Banfield, *The Unheavenly City* (Boston: Little, Brown, 1970).

43. John J. Appel, "American Negro and the Immigrant Experience: Similarities and Differences," *American Quarterly*, 18 (Spring 1966).

44. Saloutos, *Exodus—U.S.A.*

45. Fried, "Deprivation and Migration," pp. 123–127, 147–149; Carpenter, "Immigrants and Their Children"; Hutchinson, *Immigrants and Their Children, 1850–1950.*

46. Scheiner, *Negro Mecca*; Franklin E. Frazier, *The Negro in the United States* (New York: Macmillan, 1957); Gunnar Myrdal, *An American Dilemma: The Negro Problem and Modern Democracy* (New York: Harper, 1944); *The New York Times*, March 21, 1937.

47. Greer, "Attitudes Toward the Negro in New York City," pp. 45–54; Gilbert Osofsky, *Harlem: The Making of a Ghetto, Negro New York, 1890–1930* (New York: Harper and Row, 1965), pp. 18–32; Allan Spear, *Black Chicago: The Making of a Negro Ghetto, 1890–1920* (Chicago: University of Chicago Press, 1967), p. 72.

48. Glazer and Moynihan, *Beyond the Melting Pot*; Handlin, "The Goals of Integration." See: "The Negro American," *Daedalus*, 94 (Spring 1965) and 95 (Winter 1966).

49. Nathan Glazer, "Negroes and Jews: The New Challenge to Pluralism," *Commentary*, 6 (December 1964): 3.

CHAPTER 6

1. George D. Strayer, "The Survey Staff" in "School Survey: Lynn, Massachusetts" (New York: Institute of Educational Research, Teachers College, Columbia University, 1927). *See other surveys*

by Strayer: "Baltimore, Maryland" (1921), "Lancaster, Pennsylvania" (1924), "Fort Lee, New Jersey" (1927), "Missouri" (1929), "Holyoke, Massachusetts" (1930), "Hartford, Connecticut" (1937), "New York City" (1940), "Boston, Massachusetts" (1944). *See studies by other surveyists including*: James H. Van Sickle, "Brookline, Massachusetts" (1917); William H. Allen, "Cooperative and Constructive Survey—The Allen Survey: New York City Schools" (New York: New York City Board of Education, 1924); Osman R. Hull, "Los Angeles" (1934); George A. Works, "Philadelphia Public School Study" (Philadelphia Board of Education, 1937). *See also*: National Council of Education, "Standards and Tests for Measuring the Efficiency of Schools or Systems of Schools," U.S. Bureau of Education Bulletin No. 13 (1913); Henry Lester Smith and Edgar Alvin O'Dell, "A Bibliography of School Surveys and References on School Surveys," Vol. 8 (Indiana University, September-November, 1931); and "A Supplement to a Bibliography of School Surveys and References on School Surveys," Vol. 14 (Indiana University, June 1938). Together more than 200 pages about 2,000 studies at all school levels across the nation; David Siegel, "Survey and Trend Studies," *Review of Educational Research,* 12 (December, 1942); Raymond Callahan, *Education and the Cult of Efficiency* (Chicago: University of Chicago Press, 1962); David Tyack, "City Schools at the Turn of the Century: Centralization and Social Control" (Stanford, Calif.: School of Education, Stanford University, 1969; Mimeographed).

It should be pointed out that John Dewey was opposed to such surveying, complaining that it had become a convenient means of classifying and standardizing students—"the same old education masquerading in the terminology of science." National Society for the Study of Education, *The Twelfth Yearbook* (Washington, D.C., 1913).

2. Frank Jennings, "Editorial," *Teachers College Record,* 72 (September 1970).

3. For example: general data as in *Biennial Survey of Education in the U.S.: Statistics of Public Schools, 1936–37* (Washington, D.C.: U.S. Office of Education, 1937); *Biennial Survey of Education in the U.S.: Statistics of Public Schools,* 1954–55 (Washington, D.C.: U.S. Office of Education, 1955); George D. Strayer, "Age and Census of Schools and Colleges: A Study of Retardation and Elimination," U.S. Bureau of Education Bulletin No. 5 (1911).

Small-town data as available in such studies referred to in footnote 1 above. I will make reference to specific studies in major cities as I proceed.

4. Joseph Mayer Rice, *The Public School System of the United States* (New York, 1893); William R. Harper, "Report of the Educational Commission of the City of Chicago" (Chicago, 1898), pp. 2, 66, 89, 177.

5. Harper, "Report of the Educational Commission of the City of

Chicago," pp. 66, 89; Connecticut State Board of Education, "Attendance and Child Labor" (1905), and "A Study of the Costs of Secondary Education in Connecticut" (1924); Educational Commission of Cleveland, "Report on Cleveland Schools" (Cleveland Board of Education, 1906), pp. 3, 17, 21–27, 42, 71; Strayer, "Age and Census of Schools and Colleges"; Leonard Ayres, *The Identification of the Misfit Child* (New York: Russell Sage Foundation, 1911); Leonard Ayres, "Significant Development in Educational Surveying," National Education Association Journal of Proceedings and Addresses (1916), p. 92; Leonard Ayres, *Survey of School Surveys* (New York: Russell Sage Foundation, 1918); Leonard Ayres, *An Index of State School Systems* (New York: Russell Sage Foundation, 1920); Committee on Schools, Fire, Police and Civil Service of the City of Chicago, "Recommendations for Reorganization of the Public School System of the City of Chicago" (1916), pp. 1, 9, 75, 79–83; National Society for the Study of Education, "Junior High School," The Fifteenth Yearbook, Part 3 (Washington, D.C., 1916), pp. 65–82, and "Urban School Problems," The Sixteenth Yearbook (Washington, D.C., 1917), p. 186–189; Colin Greer, "Attitudes Toward the Negro in New York City 1890–1914" (Unpublished Master's Thesis, London University, 1968), pp. 45–50.

6. Committee on Schools, Fire, Police and Civil Service of the City Council of Chicago, "Recommendations for Reorganization of the Public School System of the City of Chicago," pp. 9–83; Greer, "Attitudes Toward the Negro in New York City"; Hanus, "School Inquiry of New York City Schools"; Harper, "Report of the Educational Commission of the City of Chicago," p. 89; Pennsylvania Department of Public Instruction, "Report of the Survey of the Public Schools of Philadelphia" (Public Education and Child Labor Association of Pennsylvania, 1922), Vol. 1, pp. 126–127, 221–231, 240–249; William L. Ettinger, "Survey of the Junior High Schools of the City of New York" (New York City Board of Education, 1923); George S. Counts, *Schools and Society in Chicago* (New York: Harcourt, Brace, 1928), pp. 137–139; F. P. Graves, "Report of a Study of New York City Schools" (Albany, 1933), pp. 20–43.

7. Committee on Schools, Fire, Police and Civil Service, "Recommendations for Reorganization of the Public School System of the City of Chicago," p. 75; George D. Strayer, "Report of the Survey of the Public School System of Baltimore, Md." (New York: Institute for Educational Research, Teachers College, Columbia University, 1922), pp. 30, 33; Boston Board of Education, "Survey of Boston Schools" (Boston, 1924), pp. 176–178; George D. Strayer, "Report of New York City Schools: Interim Report" (Albany, 1943), pp. 15, 322–442.

Graves, "Report of a Survey of New York City Schools," pp. 20–43; Detroit Board of Education, "Research Bulletins," No. 2 and No. 3 (Detroit, 1920), pp. 3–6; Arthur B. Moelhman, "Public

Education in Detroit" (Detroit, 1925), pp. 113, 160–168; George A. Works, "Philadelphia Public School Survey" (Philadelphia Board of Education, 1937), pp. 4–22, 143–146; Basil Coleman, *The Coleman Report: Statistics of Public Schools in the United States, 1937–1938* (Washington, D.C.: Government Printing Office, 1938), pp. 60–62.

Leonard Ayres, *The Education Survey* (Cleveland: The Cleveland Foundation, 1916), pp. 8–13; Raymond Moley, *The Education Survey Six Years After* (Cleveland: The Cleveland Foundation, 1923), p. 9; George D. Strayer, Report of the Public School System of New York City: "Interim Report" pp. 15, 220–322; George Strayer, "Report of a Survey of the Public Schools of Boston" (1944), pp. 533–556; Subcommittee of the Joint Legislative Committee to Investigate Procedures and Methods of Allocating State Monies for Public School Purposes and Subversive Activities, "Report" (1944).

8. George S. Counts, *The Selective Character of American Education* (Chicago: University of Chicago Press, 1922); Lloyd W. Warner, Robert J. Havighurst, Martin B. Loeb, *Who Shall Be Educated: The Challenge of Unequal Opportunities* (New York: Harper, 1944); Sarane S. Boocock, "Toward a Sociology of Learning: A Selective Review of Existing Research," *Sociology of Education*, 39 (Winter 1966); Christopher Jencks and David Riesman, "On Class in America," *Public Interest*, 10 (Winter 1968); Daniel Schreiber, ed., *Profile of the School Dropout*, Part II (New York: Vintage, 1968); A. J. Jaffe and Walter Adams, "American Higher Education in Transition" (New York: Bureau of Applied Social Research, Columbia University, 1969), pp. 159–167, 188–189.

9. See my chapter 5.

Frank Ross, *School Attendance in 1920* (Washington, D.C.: Government Printing Office, 1924), pp. 2–4, 35, 41, 75, 183; Niles Carpenter, "Immigrants and Their Children, 1920," Census Monograph No. 7 (Washington, D.C.: Government Printing Office, 1927), p. 32; Nathan Glazer and Daniel P. Moynihan, *Beyond the Melting Pot* (Cambridge, Mass.: M.I.T. Press, 1963), pp. 222, 224.

10. Raymond E. Callahan, *Education and the Cult of Efficiency* (Chicago: University of Chicago Press, 1962), pp. 82–100; Paul R. Mort and Francis G. Cornell, *American Schools in Transition: How Our Schools Adapt Their Practices to Changing Needs* (New York: Bureau of Publications, Teachers College, Columbia University Press, 1941), pp. 4–16.

11. Committee on Schools, Fire, Police and Civil Service, "Recommendations for Reorganization of the Public School System of the City of Chicago," pp. 75–83.

12. Lawrence Cremin, *The Transformation of the School* (New York: Knopf, 1961), Preface.

13. Mort and Cornell, *American Schools in Transition*, pp. 25–30; George D. Strayer, "Retrospective and Prospective" (New York:

Institute for Educational Research, Teachers College, Columbia University, 1931), p. 32; George D. Strayer, "Progress in City School Administration During the Past 25 Years" in Alvine Eurich, ed., *The Changing Educational World 1905–1930* (Madison, Wisc.: University of Wisconsin Press, 1931), pp. 136–161; George D. Strayer, "Planning for School Surveys" (New York: Institute for Educational Research, Teachers College, Columbia University, 1948), p. 52.

14. Mort and Cornell, *American Schools in Transition*, pp. 72–80.

15. John Dewey, "Current Problems in Secondary Education," *School Review*, 10 (1902) and 11 (1903); John Dewey, "Proceedings," National Education Association, *Journal of Proceedings and Addresses* (1912), p. 90; John Dewey, *The Sources of a Science of Education* (New York, 1929); Cremin, *Transformation of the School*; Merle Curti, *The Social Ideas of American Educators* (Toronto: Littlefield, Adams, 1961); Clarence Karier, "Testing for Order and Control in the Corporate Liberal State" (Urbana, Ill.: University of Illinois, 1971; Mimeographed), pp. 13, 19–22.

16. Coleman, *The Coleman Report*; Graves, "Report of a Study of the New York City Schools"; Moelhman, "Public Education in Detroit"; Moley, *Education Survey Six Years After*, pp. 8–13; Pennsylvania Department of Public Instruction, "Survey of Philadelphia Schools, 1922"; Boston Board of Education, "Survey of Boston Schools, 1924"; Strayer, "Survey of Chicago Schools, 1932," Vol. 1, pp. 3–4, 147–149, 153–167; Vol. 2, pp. 75, 102, 120; George D. Strayer, "The Report of a Survey of the Public Schools of the District of Columbia" (Subcommittee on D.C. Appropriations of the Senate and House of Representatives, 1949), pp. 43, 301–328, 543–555, 645–670.

17. Coleman, *The Coleman Report*, p. 121; Counts, *Schools and Society in Chicago*, pp. 137–139; Graves, "Report of a Study of the New York City Schools"; Moelhman, "Public Education in Detroit," pp. 8–13; Moley, *Education Survey Six Years After*, pp. 113, 160–168; Strayer, "Report of a Survey of the Public Schools of the District of Columbia," p. 43; Pennsylvania Department of Public Instruction, "Survey of Philadelphia Schools, 1922," pp. 133–231; Boston Board of Education, "Survey of Boston Schools, 1924"; Chicago Board of Education, "Chicago Public Schools" (1924), pp. 3, 23; Strayer, "Survey of Chicago Schools, 1932"; *Biennial Survey of Education, 1936–1938*, pp. 8–12, 60–62; *Biennial Survey of Education*, 1938–1940, pp. 50–53.

18. Pennsylvania Department of Public Instruction, "Report of the Survey of the Public Schools of Philadelphia, 1922," pp. 239–249.

19. *Biennial Survey of Education, 1936–1938*, pp. 8–12.

20. Strayer, "Survey of Boston Schools, 1944," "Survey of New York City Schools, 1944," "Survey of District of Columbia Schools, 1949," p. 43.

21. Strayer, "Survey of Boston Schools, 1944," pp. 535–539, and "Survey of New York City Schools, 1944," pp. 322–334.
22. Robert J. Havighurst, "The Public Schools of Chicago: A Survey Report" (Chicago Board of Education, 1964); James Coleman *et al, Equality of Educational Opportunity* (Washington, D.C.: Government Printing Office, 1966); Harry Passow, *Toward Creating a Model Urban School System: A Study of the Washington, D.C. Public Schools* (New York: Teachers College, Columbia University, 1967).
23. *Biennial Survey of Education,* 1936–1938, pp. 60–62. Fritz Machlup, "Comments on Human and Social Benefits of Universal University Attendance," Paper presented to 54th Annual Meeting, American Council on Education, October 7, 1971; Arthur M. Cohen, *Dateline '79: Heretical Concepts for the Community College* (Beverly Hills, Calif.: Glencoe, 1969); *New York Times,* November 17, 1971.

   After the spurt in absolute college attendance following open admission programs as they have been explored in California and New York, and the nationwide explosion of junior colleges, more than 40% of high school graduates did not go on to college. In New York, the dropout rates for open admissions students in their freshman year maintained the historic percentage of those failing, now in newly determined academic terminal points.
24. U.S. Bureau of the Census, Current Population Reports, Series P-20, No. 220, "Ethnic Origins and Educational Attainment: November, 1969" (1970), pp. 7–8, and No. 207, "Educational Attainment: March 1970" (1971), pp. 11–18.
25. Jaffe and Adams, *American Higher Education in Transition,* pp. 159–167.
26. Cremin, *Transformation of the School;* Jacob Riis, *A Ten Years' War: An Account of the Battle with the Slum in New York* (New York: Macmillan, 1902); Ellwood P. Cubberley, *Public Education in the United States* (Boston: Houghton, Mifflin, 1919).
27. New York City Board of Education, "Annual Report" for: 1900, pp. 60–69, 1901, pp. 122–160, 1904, p. 1355, 1909, pp. 148, 166, 1911, p. 166, 1915, pp. 48–49; L. E. Palmer, "New York's Truancy Problem," *Charities,* 17 (February 1906); Paul M. Hanus, "School Inquiry of New York Public Schools," Contributing Reports and Final Report (New York: Board of Estimate and Apportionment, 1913).
28. Children's Aid Society, "Annual Report," 1900, p. 42, and 1901, p. 36; A. H. Shaw, "The True Character of New York Public Schools," *World's Work,* 7 (1903–1904); Grace Abbott, *Immigrant and Community* (New York: Century, 1917); New York City Board of Education, "Fiftieth Annual Report," 1948, p. 171; Selma R. Berrol, "Schools of New York in Transition, 1898–1914," *Urban Review,* 1 (December 1966).

29. Riis, *Ten Years' War*; Rice, *Public School System of the United States*; Jacob A. Riis, "Playgrounds for City Schools, *Century*, 48 (1894); New York City Board of Education, "Annual Report of the Superintendent of Schools" for: 1901, pp. 23, 60, 1906, p. 148, 1915, pp. 48–49, 1948, pp. 62–180; Shaw, "True Character of New York Public Schools"; A. Emerson Palmer, *The New York Public School* (New York: Macmillan, 1905).

30. Committee on Congestion, "The True Story of the Worst Congestion in Any Civilized City" (New York, 1910), pp. 6–10.

31. *Ibid.*

32. New York City Board of Education, "Annual Report of the Superintendent of Schools" for: 1904, 1913, 1915, 1948; *School and Society*, 12 (1920); Hanus, "School Inquiry of New York Public Schools"; Greer, "Attitudes Toward the Negro in New York City."

33. *School Journal*, 80 (1913); New York City Board of Education, "Annual Report of the Superintendent of Schools" for 1915, pp. 48–49.

34. New York City Board of Education, "Annual Report of the Superintendent of Schools" for 1901, 1915, 1948; *Educational Review*, 38 (1908); *School Journal*, 80 (1913): 85.

35. New York City Board of Education, "Annual Report of the Superintendent of Schools," 1907, pp. 122, 131.

36. New York City Board of Education, "Annual Report of the Superintendent of Schools," 1901, 1902, 1903, 1904, 1905, 1906, 1907, 1908, 1909, 1915, 1948; Cremin, *Transformation of the School*; Greer, "Attitudes Toward the Negro in New York City"; E. G. Hartmann, *The Movement to Americanize the Immigrant* (New York: AMS Press, 1948); John Higham, *Strangers in the Land: Patterns of American Nativism 1860–1925* (New York: Atheneum, 1963); Roy Lubove, *The Professional Altruist: The Emergence of Social Work as a Career 1880–1930* (Cambridge, Mass.: Harvard University Press, 1965).

37. Hanus, "School Inquiry of New York City Public Schools"; W. M. Maxwell, *A Quarter Century of Public School Development* (New York, 1912); U.S. Bureau of the Census, "Historical Statistics of the United States, Colonial Times to 1957" (Washington, D.C., 1960), pp. 207–210, 214.

38. Hanus, "School Inquiry of New York City Public Schools"; Maxwell, *Quarter Century of Public School Development*.

39. A. S. Draper, *Our Children, Our Schools, and Our Industries* (Albany: New York State Department of Education, 1908); H. E. Miles, "Industrial Education: Report of the Committee on Industrial Education" (New York: 20th Annual Convention, National Association of Manufacturers, 1915); New York City Board of Education, "Annual Report of the Superintendent of Schools," 1915; Elizabeth Irwin, "Study of New York City Truants, 1915," quoted in Crime Commission of New York State, "Study of 201

Truants in New York City Schools" (Commission on Causes and Effects of Crime, 1927). J. K. Vandenburg, "Causes of the Elimination of Students in Public Secondary Schools of New York City," No. 47, "Contributions to Education" (New York: Teachers College, Columbia University, 1912).

40. Ayres, *Survey of School Surveys*; Leonard Ayres, "Laggards in Our Schools," in Eugene A. Nifennecker, *"Language in Our Schools"* (New York City Board of Education, 1922); David B. Cohen, "Immigrants in the Schools," *Review of Educational Research*, 40 (February 1970).

41. Carpenter, "Immigrants and Their Children," p. 132; Cohen, "Immigrants in Schools"; Graves, "Study of New York City Schools," pp. 20–43; Moley, *Education Study Six Years After*; Mary Antin, *The Promised Land* (New York: Houghton, Mifflin, 1912); Detroit Board of Education, "Age, Grade and Nationality Survey," *The Annals of the American Academy of Political and Social Science*, 98 (1921); Walter Laidlaw, *Statistical Sources for Demographic Studies, Greater New York, 1920* (New York: New York Federation of Churches, 1922), p. 135; *Mental Hygiene*, 7 (1923).

42. Cohen, "Immigrants in Schools."

43. W. L. Ettinger, "Ten Addresses" (New York City Board of Education, 1923).

44. Crime Commission of New York State, "Study of 201 Truants in New York City Schools."

45. *The New York Times*, May 15, 1938; December 14, 1940.

46. Graves, "Survey of New York City Schools," pp. 20–43; New York City Board of Education, "Annual Report of the Superintendent of Schools," 1934, p. 166, and "Special Report: Handicapped and Underprivileged Children 1934"; *The New York Times*, August 15, 1930; June 28, 1931; December 17, 1931; May 4, 1933; May 21, 1935; March 4, 1936; October 28, 1936; May 15, 1938, II; June 4, 1939, II; September 9, 1939; September 13, 1939; October 8, 1939, VI.

47. *The New York Times*, December 30, 1934; May 15, 1938.

48. New York City Board of Education, "Annual Report of the Superintendent of Schools," 1934, 1937, 1948, p. 153.

49. J. B. Maller, "The Economic and Social Correlatives of School Progress in New York City," *Teachers College Record* (May 1933): 25.

50. Maller, "Economic and Social Correlatives of School Progress"; *The New York Times*, August 27, 1931; May 4, 1933; December 30, 1934; New York City Board of Education, "Annual Report of the Superintendent of Schools," 1934, and "Special Report: Handicapped and Underprivileged Children" (1934).

51. William Jensen, "Annual Report of the Superintendent of Schools: Our Public Schools 1948–1950," (New York City Board of Education, 1950), p. 25. *The New York Times*, January 19, 1950; New York City Board of Education, "A Ten Year Report from Principals of Districts 23 and 24 (New York, 1937), and "Youth in School

and Industry" (New York, 1937); F. M. Wilson and M. Krugman, "Studies of Student Personnel" (Albany: Planning Committee of the Cooperative Study of Vocational Education in New York City Schools, 1951).

52. *The New York Times*, June 4, 1939, II; October 8, 1939, VI; April 1, 1943; June 27, 1943, IV; September 11, 1943.

53. *The New York Times*, May 21, 1935; December 1, 1935, IV; March 21, 1937, II, New York City Board of Education, Committee on Maladjustment and Delinquency, "Report on Delinquency in New York City Schools" (1937); and "The Harlem Project: The Role of the School in Preventing and Correcting Maladjustment and Delinquency" (1949).

54. *The New York Times*, December 30, 1934, VIII; December 1, 1935, IV; March 4, 1936; October 28, 1936; November 22, 1936, II; March 21, 1937, II.

55. New York City Board of Education, "Annual Report of the Superintendent of Schools," 1948, p. 162.

56. Allen, "New York City Schools, 1924"; Graves, "Survey of New York City Schools"; Strayer, "Survey of New York City Schools."

57. New York City Board of Education, "The Harlem Project"; see: *The New York Times*, September 29, 1945; October 18, 1945; October 21, 1945; December 10, 1945; December 11, 1945; August 5, 1946; January 6, 1947; February 18, 1947; February 3, 1948; October 12, 1948; June 3, 1949; November 17, 1949, for continued inability of schools to deal successfully with the needs of significant parts of the city's white school population; *New York Times*, March 12, 1950; May 23, 1950; May 16, 1954; May 16, 1959.

CHAPTER 7

1. Colin Greer, "Attitudes Toward the Negro in New York City 1890–1914" (Unpublished Master's Thesis, London University, 1968); and *Cobweb Attitudes: Essays on American Education and Culture* (New York: Teachers College Press, Columbia University, 1970).

2. W. E. B. DuBois, *The Philadelphia Negro* (Philadelphia, 1899), pp. 25–60; Gilbert Osofsky, *Harlem: The Making of a Ghetto, Negro New York 1890–1930* (New York: Harper and Row, 1965), pp. 3–20; Allan H. Spear, *Black Chicago: The Making of a Negro Ghetto 1890–1920* (Chicago: University of Chicago Press, 1967), pp. 11–25; Stephen R. Fox, *The Guardian of Boston: William Monroe Trotter* (New York: Atheneum, 1970); A. J. Jaffe and Walter Adams, "American Higher Education in Transition" (Bureau of Applied Social Research, Columbia University, 1969); U.S. Bureau of the Census, Current Population Reports, Series P-20, No. 220, "Ethnic Origin and Educational Attainment: November 1969" (1970); and Current Population Reports, Series P-20, No. 207, "Educational Attainment: March 1970" (1971).

3. U.S. Bureau of the Census, Supplementary Report, "Negro Population in Selected Places and Selected Counties," Series PC (S1) No. 2 (1971).

4. Osofsky, *Making of a Ghetto*; Seth M. Scheiner, *Negro Mecca: A History of the Negro in New York 1865-1920* (New York: New York University Press, 1965), pp. 1–21; Greer, "Attitudes Toward the Negro in New York City."

5. Osofsky, *Making of a Ghetto*; Scheiner, *Negro Mecca*.

6. Scheiner, *Negro Mecca*; Greer, "Attitudes Toward the Negro in New York City."

7. National Labor Committee, "Child Workers of the Nation" (Washington, D.C., 1909); Lillian Wald, *The House on Henry Street*, (New York: Holt, 1915); Greer, "Attitudes Toward the Negro in New York City."

8. Mary White Ovington, *Half a Man: The Status of the Negro in New York* (New York, 1911); George E. Hayres, *The Negro at Work in New York: A Study in Economic Progress* (New York, 1912), and "Conditions Among Negroes in Cities," *The Annals of the American Academy of Political and Social Science*, 49 (September, 1913); Herman D. Bloch, "The New York City Negro and Occupational Eviction 1860–1910," *International Review of Social History*, Vol. 5, Part 1 (1960).

9. Greer, "Attitudes Toward the Negro in New York City."

10. *Ibid.*

11. *Ibid.*; E. Franklin Frazier, *The Negro in the United States* (New York: Macmillan, 1957).

12. Greer, "Attitudes Toward the Negro in New York City"; Osofsky, *Making of a Ghetto*; Spear, *Black Chicago*.

13. Greer, "Attitudes Toward the Negro in New York City."

14. *Charities*, 15 (October 7, 1905).

15. U.S. Department of Labor Bulletin, "The Condition of Negroes in Various Cities" (May 1897); Frances Kellor, *Experimental Sociology* (New York, 1902), Introduction; Greer, "Attitudes Toward the Negro in New York City"; William Ryan, *Blaming the Victim* (New York: Pantheon, 1971).

16. U.S. Department of Labor Bulletin, "Condition of Negroes in Various Cities"; W. E. Chancellor, "The Education of Colored Persons in America," *School Journal*, 80 (1913); Greer, "Attitudes Toward the Negro in New York City."

17. Frances Blascoer, *Colored School Children in New York* (New York, 1915), Foreword.

18. *Ibid.*; Greer, "Attitudes Toward the Negro in New York City."

19. *Ibid.*

20. Blascoer, *Colored School Children in New York*.

21. *Ibid.*, pp. 10–15.

22. *Ibid.*, Introduction, pp. 108–120; W. H. Odum, "Negro Children in the Public Schools of Philadelphia," *The Annals of the American Academy of Political and Social Science*, 44 (September 1913).

23. Blascoer, *Colored School Children in New York*, pp. 172–175; Odum, "Negro Children in the Public Schools of Philadelphia."
24. M. J. Mayo, "The Mental Capacity of the American Negro," *Archives of Psychology*, 28 (1913), pp. 10–21.
25. *Ibid.*, pp. 23–25.
26. Blascoer, *Colored School Children in New York*; Odum, "Negro Children in the Public Schools of Philadelphia"; Mayo, "Mental Capacity of the American Negro."
27. Blascoer, *Colored School Children in New York*; Odum, "Negro Children in the Public Schools of Philadelphia"; Mayo, "Mental Capacity of the American Negro," pp. 40–45.
28. Mayo, "Mental Capacity of the American Negro."
29. Blascoer, *Colored School Children in New York*; *The Annals of the American Academy of Political and Social Science*, 98 (November, 1921): 142.
30. John Higham, *Strangers in the Land: Patterns of American Nativism 1860–1925* (New York: Atheneum, 1963).
31. New York City Board of Education, "Annual Report of the Superintendent of Schools," 1933–1934; "Special Report: Elementary Schools of New York City—Their Problems and the Efforts That Are Being Made to Solve Them" (New York, 1934).
32. Maudelle B. Bousfield, "The Intellectual and School Achievement of Negro Children," *Journal of Negro Education*, 1 (1932): 131.
33. "Complete Riot Report: Mayor's Commission to Investigate Conditions in Harlem," *New York Amsterdam News*, July 18, 1936.
34. *Ibid.*;*The New York Times*, February 23, 1936, II; March 21, 1937, II; April 29, 1938; December 28, 1939.
35. "Complete Riot Report," *New York Amsterdam News*; *The New York Times*, March 21, 1937, II; December 28, 1939; April 18, 1944; Frazier, *Negro in the United States*.
36. "White Parents Boycott Schools to Express Their Dissatisfaction with Them," *The New York Times*, February 2, 1939; March 4, 1939; October 12, 1948; June 2, 1949; March 12, 1950; March 23, 1950; "General Economic Conditions and New Developments on Schools," *The New York Times*, October 3, 1943, IV; May 31, 1944; February 25, 1944; April 18, 1944; October 18, 1945; December 10, 1945; New York City Board of Education, "The Harlem Project: The Role of the School in Preventing and Correcting Maladjustment and Delinquency" (New York, 1949).
37. *New York Times Index*.
38. New York City Board of Education, Commission on Integration, "Sub-Commission Reports" (New York, 1956); *The New York Times*, March 1, 1956; December 2, 1956, IV; February 26, 1957; February 28, 1957; May 14, 1957; May 17, 1957; September 9, 1958; September 20, 1959; New York City Board of Education, "Annual Report of the Superintendent of Schools" for 1957, 1966; Marilyn Gittell, *Participants and Participation in New York City* (New York: Center for Urban Education, 1967); Superintendent of Schools, Archdiocese of New York, "Rise in Non-Public School

Attendance in Manhattan and the Bronx" (January, 1969; Unpublished memorandum).

39. New York City Board of Education, "Annual Report of the Superintendent of Schools," 1957; State Education Commissioner's Advisory Committee on Human Relations and Community Tensions, "Desegregating the Public Schools of New York City" (New York: Institute of Urban Studies, Teachers College, Columbia University, 1964); U.S. Bureau of the Census, 1970 Census of Population, Supplementary Report, "Negro Population in Selected Places."

40. Gittell, *Participants and Participation*; Bert Swanson, *Struggle for Equality: School Integration Controversy in New York City* (New York: Hobbs, Dorman, 1966).

41. *The New York Times*, February 2, 1939; March 4, 1939; October 12, 1948; June 3, 1949; March 12, 1950; March 23, 1950; February 28, 1955; October 6, 1955, IV; May 14, 1957; Educational Policies Commission, "Higher Education in a Decade of Depression" (Washington, D.C.: National Education Association, 1957); President's Committee on Education Beyond the High School, "Second Report to the President" (Washington, D.C.: Government Printing Office, 1957); James B. Conant, *The American High School Today* (New York: McGraw-Hill, 1959).

Bibliography

BOOKS

Abbott, Grace. *Immigrant and Community*. New York: Century, 1917.

Addams, Jane. *Forty Years at Hull House*. New York: Macmillan, 1929.

Alloway, David N., and Cordasco, Francesco. *Minorities and the American City*. New York: Vintage, 1968.

Antin, Mary. *The Promised Land*. New York: Houghton, Mifflin, 1912.

Ayres, Leonard. *The Education Survey*. Cleveland: The Cleveland Foundation, 1923.

―――. *The Identification of the Misfit Child*. New York: Russell Sage Foundation, 1911.

―――. *An Index of State School Systems*. New York: Russell Sage Foundation, 1921.

―――. *Survey of School Surveys*. New York: Russell Sage Foundation, 1918.

Bailyn, Bernard. *Education in the Forming of American Society*. New York: Vintage, 1960.

Banfield, Edward. *The Unheavenly City*. Boston: Little, Brown, 1970.

Benson, Lee. *The Concept of Jacksonian Democracy*. Princeton: Princeton University Press, 1961.

Berg, Ivar. *Education and Jobs: The Great Training Robbery*. New York: Praeger, 1970.

Berkson, Isaac B. *Theories of Americanization*. New York: Teachers College, Columbia University, 1920.

Bernstein, Barton J. (ed.). *Towards a New Past: Dissenting Essays in American History*. New York: Pantheon, 1968.

Berwanger, Eugene H. *The Frontier Against Slavery: Western Anti-Negro Prejudice and the Slavery Extension Controversy*. Urbana, Ill.: University of Illinois Press, 1969.

Blascoer, Frances. *Colored School Children in New York*. New York: 1915.

Blau, Peter, and Duncan, Otis B. *The American Occupational Structure*. New York: Wiley, 1967.

Boorstin, Daniel J. *The Americans: The Colonial Experience*. New York: Vintage, 1958.

Boorstin, Daniel J. *The Americans: The National Experience*. New York: Vintage, 1965.

Broehl, Wayne G. Jr. *The Molly Maguires*. Cambridge, Mass.: Harvard University Press, 1964.

Brotz, Howard. *The Black Jews*. New York: Free Press, 1964.

Callahan, Raymond E. *Education and the Cult of Efficiency*. Chicago: University of Chicago Press, 1962.

Clark, Kenneth B. (ed.). *Racism and American Education*. New York: Harper/Colophon, 1970.

Clark, Kenneth B. *Dark Ghetto*. New York: Harper and Row, 1965.

Cleaver, Eldridge. *Soul on Ice*. New York: Ramparts, McGraw-Hill, 1968.

Cohen, Arthur M. *Dateline '79: Heretical Concepts for the Community College*. Beverly Hills, Calif.: Glencoe, 1969.

Cohen, Sol. *Progressives and Urban School Reform: The Public Education Association of New York City, 1895–1954*. New York: Teachers College Press, Columbia University, 1964.

Coleman, Basil. *The Coleman Report: Statistics of Public Schools in the United States, 1937–1938*. Washington, D.C.: Government Printing Office, 1938.

Coleman, James *et al. Equality of Educational Opportunity*. Washington, D.C.: Government Printing Office, 1966.

Coles, Robert. *Still Hungry in America*. New York: New American Library, 1969.

Commager, Henry Steele. *The American Mind*. New Haven: Yale University Press, 1950.

Committee on the Role of Education in American History. *Education and American History*. New York: Fund for the Advancement of Education, 1965.

Conant, James B. *The American High School Today*. New York: McGraw-Hill, 1959.

Conway, Alan. *The Welsh in America: Letters from the Immigrants*. Minneapolis: University of Minnesota Press, 1961.

Coons, John, *et al. Private Wealth and Public Education*. Cambridge: Harvard University Press, 1970.

Counts, George S. *Schools and Society in Chicago*. New York: Harcourt, Brace, 1928.

———. *The Selective Character of American Education*. Chicago: University of Chicago Press, 1922.

Covello, Leonard, and D'Agostino, Guido. *The Heart is the Teacher*. New York: McGraw-Hill, 1958.

Cremin, Lawrence A. *The American Common School: An Historic Concept*. New York: Teachers College Press, Columbia University, 1951.

———. *The Genius of American Education*. New York: Vintage, 1965.

——— (ed.). *The Republic and the School: Horace Mann on the Education of Free Men*. New York: Teachers College Press, Columbia University, 1957.

———. *The Transformation of the School*. New York: Knopf, 1961.

———. *The Wonderful World of Ellwood Patterson Cubberley*. New York: Teachers College Press, Columbia University, 1965.

Cronon, Edmund D. *Black Moses: The Story of Marcus Garvey and the Universal Negro Improvement Association*. Madison, Wisc.: University of Wisconsin Press, 1955.

Cruse, Harold. *The Crisis of the Negro Intellectual*. New York: Morrow, 1967.

Cubberley, Ellwood P. *Changing Conceptions of Education*. Boston: Houghton, Mifflin, 1909.

———. *Public Education in the United States*. Boston: Houghton, Mifflin, 1919.

Curti, Merle. *The Social Ideas of American Educators*. Toronto: Littlefield, Adams, 1961.

De Tocqueville, Alexis. *Democracy in America*. New York: Vintage, 1959.

Dewey, John. *The Sources of a Science of Education*. New York: 1929.

Draper, A. S. *Our Children, Our Schools, and Our Institutions*. Albany: New York State Department of Education, 1908.

DuBois, W. E. B. *The Philadelphia Negro*. Philadelphia: 1899.

Dworkin, Martin S. (ed.). *Dewey on Education*, A Centennial Review. New York: Teachers College Press, Columbia University, 1959.

Erikson, Erik H. *Young Man Luther*. New York: Norton, 1958.

Ernst, Robert. *Immigrant Life in New York City, 1825–1863*. New York: I. J. Friedman, 1965.

Fairchild, Henry Pratt (ed.). *Immigrant Backgrounds*. New York: Wiley, 1927.

Ford, Paul Leicester (ed.). *Thomas Jefferson, Works*. New York: 1904.

Fox, Stephen R. *The Guardian of Boston: William Monroe Trotter*. New York: Atheneum, 1970.

Frazier, E. Franklin. *The Negro in the United States*. New York: Macmillan, 1957.

Gittell, Marilyn. *Participants and Participation in New York City*. New York: Center for Urban Education, 1967.

Glanz, Rudolph. *Jews and Irish: Historic Group Relations and Immigration*. New York: 1966.

Glazer, Nathan, and Moynihan, Daniel P. *Beyond the Melting Pot: The Negroes, Puerto Ricans, Jews, Italians and Irish of New York City*. Cambridge, Mass.: M.I.T. Press, 1963.

Goodman, Paul. *Compulsory Mis-education*. New York: Vintage, 1964.

Gordon, Milton M. *Assimilation in American Life*. London: Oxford University Press, 1964.

Greeley, Andrew. *Why Can't They Be Like Us? America's White Ethnic Groups*. New York: Dutton, 1971.

Greene, Maxine. *The Public School and the Private Vision: A Search for America in Education and Literature*. New York: Random House, 1965.

Greer, Colin. *Cobweb Attitudes: Essays on American Education and Culture*. New York: Teachers College Press, Columbia University, 1970.

Handlin, Oscar. *The Uprooted*. New York: Grosset and Dunlap, 1951.

Hansen, Allen. *Liberalism and American Education in the Eighteenth Century*. New York: Octagon, 1965.

Hansen, Marcus L. *The Problem of the Third Generation Immigrant*. Rock Island, Ill.: Augustana Historical Society Publications, 1938.

Hartmann, E. G. *The Movement to Americanize the Immigrant*. New York: AMS Press, 1948.

Hartz, Louis. *The Founding of New Societies*. New York: Harbinger Books, 1964.

————. *The Liberal Tradition in America*. New York: Harcourt, Brace, 1955.

Havighurst, Robert J., and Neugarten, Berenice L. *Society and Education*. Boston: Allyn and Bacon, 1957.

Hawgood, John A. *The Tragedy of German-America*. New York: Putnam, 1940.

Hayres, George E. *The Negro at Work in New York: A Study in Economic Progress*. New York: 1912.

Heilbroner, Robert L. *The Future as History*. New York: Harper and Row, 1959.

Henderson, Algo G. *Policies and Practices in Higher Education*. New York: Harper and Brothers, 1960.

Herberg, Will. *Catholic-Protestant-Jew*. New York: Doubleday, 1955.

Higham, John. *Strangers in the Land: Patterns of American Nativism, 1860–1925*. New York: Atheneum, 1963.

Hofstadter, Richard. *Anti-Intellectualism in American Life*. New York: Knopf, 1966.

Holt, John. *How Children Fail*. New York: Pitman, 1965.

Hourwich, Isaac A. *Immigration and Labor*. New York: B. W. Huebsh, 1922.

Hum Lee, Rose. *The Chinese in the United States of America*. Cambridge: Oxford University Press, 1960.

Hutchinson, E. P. *Immigrants and Their Children, 1850–1950*. New York: Wiley, 1956.

Illich, Ivan. *Deschooling Society*. New York: Harper and Row, 1971.

Jencks, Christopher, and Riesman, David. *The Academic Revolution*. New York: Doubleday, 1968.

Katz, Michael B. *The Irony of Early School Reform: Educational Innovation in Mid-Nineteenth Century Massachusetts.* Cambridge, Mass.: Harvard University Press, 1968.

Kellor, Frances. *Experimental Sociology.* New York: 1902.

Kolko, Gabriel. *The Triumph of Conservatism.* New York: Free Press, 1963.

————. *Wealth and Power in America: An Analysis of Social Class and Income Distribution.* New York: Praeger, 1962.

Kraditor, Eileen S. *Means and Ends in American Abolitionism: Garrison and His Critics, 1834–1850.* New York: Pantheon, 1969.

Krug, Edward A. *The Shaping of the American High School, 1880–1920.* Madison, Wisc.: University of Wisconsin Press, 1969.

Labarce, Leonard *et al. The Papers of Benjamin Franklin.* New Haven: Yale University Press, 1961.

Lanzer, William. *Political and Social Upheaval, 1832–1852.* New York: Harper and Row, 1969.

Lenski, Gerhard. *The Religious Factor.* Garden City, New York: Doubleday, 1961.

Lieberson, Stanley. *Ethnic Patterns in American Cities.* New York: Free Press, 1963.

Lipset, Seymour, and Bendix, Reinhold. *Social Mobility in Industrial Society.* Berkeley, Calif.: University of California Press, 1969.

Lowi, Theodore. *At the Pleasure of the Mayor: Patronage and Power in New York City, 1898–1958.* New York: Free Press, 1964.

Lubove, Roy. *The Professional Altruist: The Emergence of Social Work as a Career, 1880–1930.* Cambridge, Mass.: Harvard University Press, 1965.

Mayer, Martin. *The Schools.* New York: Harper and Row, 1961.

Maxwell, W. M. *A Quarter Century of Public School Development.* New York: 1912.

McCluskey, Neil G. (ed.). *Catholic Education in America: A Documentary History.* New York: Teachers College Press, Columbia University, 1966.

Meyers, Marvin. *The Jacksonian Persuasion: Politics and Belief.* Stanford, Calif.: Stanford University Press, 1957.

Middlekauff, Robert. *Ancients and Axioms: Secondary Education in Eighteenth Century New England.* New Haven: Yale University Press, 1963.

Miller, S. M., and Roby, Pamela. *The Future of Inequality.* New York: Basic Books, 1970.

Moley, Raymond. *The Education Survey Six Years After.* Cleveland: The Cleveland Foundation, 1923.

Moore, Barrington. *Social Origins of Dictatorship and Democracy: Lord and Peasant in the Making of the Modern World.* Boston: Beacon, 1966.

Mort, Paul R., and Cornell, Francis G. *American Schools in Transition: How Our Schools Adapt their Practices to Changing Needs.* New

York: Bureau of Publications, Teachers College, Columbia University, 1941.

Muggeridge, Kitty. *Beatrice Webb: A Life, 1858–1943.* London: Secker and Warburg, 1967.

Myrdal, Gunnar. *An American Dilemma: The Negro Problem and Modern Democracy.* New York: Harper, 1944.

National Urban Coalition. *Counterbudget: A Blueprint for Changing National Priorities.* New York: Praeger, 1971.

Osofsky, Gilbert. *Harlem: The Making of a Ghetto Negro New York, 1890–1930.* New York: Harper and Row, 1965.

Ovington, Mary White. *Half a Man: The Status of the Negro in New York.* New York: 1911.

Palmer, A. Emerson. *The New York Public School.* New York: Macmillan, 1905.

Park, Robert E. *The Immigrant Press and Its Control.* New York: Harper, 1922.

———, and Miller, Herbert A. *Old World Traits Transplanted.* New York: Harper, 1921.

Passow, Harry. *Toward Creating a Model Urban School System: A Study of the Washington, D.C. Public Schools.* New York: Teachers College, Columbia University, 1967.

Perkinson, Henry J. *The Imperfect Panacea: American Faith in Education, 1865–1965.* New York: Random, 1968.

Phenix, Philip H. *Realms of Meaning.* New York: McGraw-Hill, 1964.

Rein, Martin. *Social Policy.* New York: Random House, 1970.

Rice, Joseph Mayer. *The Public School System of the United States.* New York: 1893.

Riis, Jacob. *How the Other Half Lives: Studies Among the Tenements of New York.* New York: 1890; Sagamore Press, 1957.

———. *A Ten Years' War: An Account of the Battle with the Slum in New York.* New York: Macmillan, 1902.

Rischin, Moses. *The Promised City: New York's Jews, 1870–1914.* Cambridge, Mass.: Harvard University Press, 1962.

Rogers, David. *110 Livingston Street.* New York: Random House, 1968.

Rosenthal, Robert, and Jacobson, Lenore. *Pygmalion in the Classroom.* New York: Holt, Rinehart, and Winston, 1968.

Rothman, David J. *The Discovery of the Asylum: Social Order and Disorder in the New Republic.* Boston: Little, Brown, 1971.

Rude, George. *The Crowd in the French Revolution.* Oxford: Clarendon Press, 1967.

Ryan, William. *Blaming the Victim.* New York: Pantheon, 1971.

Saloutos, Theodore. *The Greeks in the United States.* New York: Teachers College Press, Columbia University, 1967.

Scheiner, Seth M. *Negro Mecca: A History of the Negro in New York City, 1865–1920.* New York: New York University Press, 1965.

Schlesinger, Arthur, Jr. *Political and Social History of the United*

States, 1829–1924. New York: Macmillan, 1951.

Schrag, Peter. *The Decline of the WASP.* New York: Random House, 1971.

―――. *Voices in the Classrooms: Public Schools and Public Attitudes.* Boston: Beacon Press, 1965.

Schreiber, Daniel (ed.). *Profile of the School Dropout.* New York: Vintage, 1968.

Sexton, Patricia Cayo. *The American School: A Sociological Analysis.* New York: Viking, 1961.

―――. *Education and Income: Inequalities in Our Public Schools.* Englewood Cliffs, N.J.: Prentice-Hall, 1961.

Shannon, William. *The American Irish.* New York: Collier-Macmillan, 1963.

Silberman, Charles. *Crisis in Black and White.* New York: Random House, 1964.

Simirenko, Alex. *Pilgrims, Colonists and Frontiersmen: An Ethnic Community in Transition.* New York: Free Press, 1964.

Sizer, Theodore R. *Secondary Schools at the Turn of the Century.* New Haven: Yale University Press, 1964.

Spear, Allan H. *Black Chicago: The Making of a Negro Ghetto, 1890–1920.* Chicago: University of Chicago Press, 1967.

Stewart, W. A. C. *The Educational Innovators: Progressive Schools, 1881–1967.* London: Macmillan, 1968.

Storr, Richard J. *The Role of Education in American History.* New York: Fund for the Advancement of Education, 1957.

Strickland, Charles E., and Burgess, Charles (ed.). *Health, Growth, and Heredity: G. Stanley Hall on Natural Education.* New York: Teachers College Press, Columbia University, 1965.

Swanson, Bert. *Struggle for Equality: School Integration Controversy in New York City* (New York: Hobbs, Dorman, 1966).

Thomas, W. I., and Znaniecki, Florian. *The Polish Peasant in Europe and America.* Chicago: University of Chicago Press, 1918; New York: Knopf, 1928; Dover, 1958.

Trent, James, and Medsher, Leland. *Beyond High School.* San Francisco: Jossey-Bass, 1966.

Turner, Frederick Jackson. *The Frontier in American History.* New York: Holt, Rinehart, and Winston, 1962.

Tyack, David B. (ed.). *Turning Points in American Educational History.* Waltham, Mass.: Blaisdell, 1967.

U.S. Riot Commission. *Report of the National Advisory Committee on Civil Disorders.* New York: Bantam Books, 1968.

Voegeli, V. J. *Free But Not Equal: The Midwest and the Negro During the Civil War.* Chicago: University of Chicago Press, 1969.

Wald, Lillian. *The House on Henry Street.* New York: Holt, 1915.

Havighurst, Robert J., and Loeb, Martin B. *Who Shall Be Educated: The Challenge of Unequal Opportunities.* New York: Harper, 1944.

Warner, Lloyd W., and Srole, Leo. *The Social Systems of American Ethnic Groups.* New Haven: Yale University Press, 1964.

Webster, Daniel. *Works Vol. 1.* Boston: Little, Brown, 1954.

Weinstein, James. *The Corporate Ideal in the Liberal State, 1900–1918.* Boston: Beacon, 1968.

Wells, H. G. *The New Machiavellians.* Garden City, N.Y.: Doubleday, Duncan, 1932.

Welter, Rush. *Popular Education and Democratic Thought in America.* New York: Columbia University Press, 1962.

Williams, William Appleman. *The Tragedy of American Diplomacy.* New York: Dell, 1970.

Woodward, C. Vann. *The Burden of Southern History.* New York: Vintage, 1960.

Znaniecki, Florian. *The Polish Peasant in Europe and America.* Chicago: University of Chicago Press, 1918; New York: Knopf, 1928.

ARTICLES, REPORTS, ETC.

Addams, Jane. "The Public School and the Immigrant Child," in National Education Association, *Journal of Proceedings and Addresses* (1908).

Anderson, Perry. "Components of the National Culture," in A. Cockburn and R. Blackburn (eds.), *Student Power,* Hammondsworth, England: Penguin Books, 1969.

Appel, John J. "American Negro and the Immigrant Experience: Similarities and Differences," *American Quarterly,* 18 (Spring 1966).

Aptheker, Herbert (ed.). "Some Unpublished Writings of W. E. B. DuBois," *Freedomways,* 5 (Winter, 1965).

Ayres, Leonard. "Laggards in Our Schools," in Eugene A. Nifennecker, *Language in Our Schools,* New York: New York City Board of Education, 1922.

———. "Significant Developments in Educational Surveying," National Education Association, *Journal of Proceedings and Addresses* (1916).

Bailyn, Bernard. "Some Historical Notes," in John Watton and James L. Kueth (eds.), *The Discipline of Education,* Madison, Wisc.: University of Wisconsin Press, 1963.

Berrol, Selma R. "Schools of New York in Transition, 1898–1914," *Urban Review,* 1 (December 1966).

Bloch, Herman D. "The New York City Negro and Occupational Eviction, 1860–1910," *International Review of Social History,* Vol. 5, Part 1 (1960).

Bøhn, Tora. "A Quest for Norwegian Folk Art in America," *Norwegian American Studies and Records,* 19 (1956).

Boocock, Sarane S. "Toward a Sociology of Learning: A Selective Review of Existing Research," *Sociology of Education,* 39 (Winter 1966).

Bousfield, Maudelle B. "The Intellectual and School Achievement of Negro Children," *Journal of Negro Education,* 1 (1932).

Brickman, William W. "Revisionism and the Study of the History of Education," *History of Education Quarterly*, 4 (1964).

Burgess, Charles. "The Educational State in America" (Unpublished Doctoral Dissertation, 1962).

Carpenter, Niles. "Immigrants and Their Children, 1920," Census Monograph No. 7, Washington, D.C.: Government Printing Office, 1927.

Cestello, Bosco D. "Catholics in American Commerce and Industry, 1925–1945," *The American Catholic Sociological Review*, 17 (October 1956).

Chancellor, W. E. "The Education of Colored Persons in America," *School Journal*, 80 (1913).

Children's Aid Society. "Annual Report," 1900.

Claghorn, Kate Haliday. "First Year's Work of a New State Bureau," *Survey*, 15 (May 11, 1912).

———. "Immigration and Its Relation to Pauperism," *The Annals of the American Academy of Political and Social Science*, 24 (October 1904).

Cohen, David. "Immigrants in the Schools," *Review of Educational Research*, 40 (February, 1970).

Committee on Congestion. "The True Story of the Worst Congestion in Any Civilized City," New York: 1910.

"Complete Riot Report: Mayor's Commission to Investigate Conditions in Harlem," *New York Amsterdam News* (July 18, 1936).

Cremin, Lawrence A. "Review of Bernard Bailyn's *Education in the Forming of American Society*," *History of Education Quarterly*, 4 (1964).

Crime Commission of New York State. "Study of 201 Truants in New York City Schools," Commission on Causes and Effects of Crime, 1927.

Curti, Merle. "The Impact of the Revolutions of 1848 on American Thought," *Proceedings of the American Philosophical Society*, 93 (June 1949).

Dewey, John. "Current Problems in Secondary Education," *School Review*, 10 (1902) and 11 (1903).

———. "Proceedings," in National Education Association, *Journal of Proceedings and Addresses* (1912).

Dodson, Dan. "Education and the Powerless," in A. Harry Passow et al., *Education of the Disadvantaged*, New York: Holt, Rinehart, and Winston, 1967.

Dowie, J. Iverne. "The American Negro on a New Frontier," in O. F. Ander (ed.), *In the Trek of the Immigrant*, Rock Island, Ill.: Augustana College Library, 1964.

Educational Policies Commission. "Higher Education in a Decade of Depression," Washington, D.C.: National Education Association, 1957.

———. "Public Education and the Future of America," Washington,

D.C.: National Education Association and American Association of School Administrators, 1955.

Elson, Ruth Miller. "American Schoolbooks and Culture in the Nineteenth Century," *Mississippi Valley Historical Review*, 46 (December 1959).

———. "Immigrants and Schoolbooks in the Nineteenth Century," History of Education Society Eastern Regional Meeting, April 1971.

Ettinger, W. L. "Ten Addresses," New York: New York City Board of Education, 1923.

Fellows, Thomas. "Educating Young Blacks," in *The New York Times* (May 14, 1971), Letter to the Editor.

Fergusson, Ernest, in *The Baltimore Sun* (May 6, 1969).

"The Foreign Element in New York City," *Harper's Magazine* (1890).

Fried, Marc. "Deprivation and Migration: Dilemmas of Causal Interpretation," in Daniel P. Moynihan (ed.), *On Understanding Poverty*, New York: Basic Books, 1969.

Glazer, Nathan. "Ethnic Groups in America: From National Culture to Ideology," in M. Berger, T. Abel, and C. H. Page (eds.), *Freedom and Control in Modern Society*, New York: Octagon, 1964.

———. "Negroes and Jews: The New Challenge to Pluralism," *Commentary*, 38 (December 1964).

Goldberg, Miriam L. "Problems in the Evaluation of Compensatory Programs for Disadvantaged Children," *Journal of School Psychology*, 4 (Spring 1966).

Greer, Colin. "Attitudes Toward the Negro in New York City, 1890–1914" (Unpublished Master's Thesis, London University, 1968).

———. "Immigrants, Negroes, and the Public Schools," *The Urban Review*, 3 (January, 1969).

———. "Public Schools: Myth of the Melting Pot," *Saturday Review* (November 15, 1969).

———. "A View From Coney Island," *The Center Forum*, 2 (December 20, 1967).

Haglund, A. William. "Finnish Immigrant Farmers in New York, 1910–1960," in O. F. Ander (ed.), *In the Trek of the Immigrant*, Rock Island, Ill.: Augustana College Library, 1964.

Hamill, Pete. "The White High Schools," *New York Post* (March 11, 1970).

Handlin, Oscar. "The Goals of Integration," *Daedalus*, 95 (Winter 1966).

———, and Handlin, Mary. "Mobility," in Edward N. Saveth (ed.), *American History and the Social Sciences*, New York: Free Press, 1964).

Hartz, Louis. "The Reactionary Enlightenment," *Western Political Quarterly*, 5 (March, 1952).

Hayres, George E. "Conditions Among Negroes in Cities," *The Annals of the American Academy of Political and Social Science*, 49 (September 1913).

Hechinger, Fred. "CUNY Begins Vital Test of Open Admissions,"
  The New York Times (September 20, 1970).

———. "School Vouchers: Can the Plan Work?" *The New York
  Times* (June 7, 1970).

Horlick, Allan. "Counting Houses and Clerks: A Study of the Social
  Control of Young Men in New York, 1840–1950" (Unpublished

"Idea of the English School, Sketch'd Out for the Consideration of the
  Trustees of the Philadelphia Academy" in Leonard Labarce *et al*
  (eds.), *The Papers of Benjamin Franklin* (New Haven: Yale University Press, 1961).

  Doctoral Dissertation, University of Wisconsin, 1969.

"Immigrants and Schoolbooks in the Nineteenth Century," History of
  Education Society, Eastern Regional Meeting (April 1971).

Iwata, Masakazo. "The Japanese Immigrants in California Agriculture,"
  Agricultural History (January 1962).

Jaffe, A. J., and Adams, Walter. "American Higher Education in
  Transition," Bureau of Applied Social Research, Columbia University, 1969.

Jencks, Christopher, and Riesman, David. "On Class in America,"
  Public Interest, 10 (Winter 1968).

Jennings, Frank. "Editorial," *Teachers College Record*, 72 (September
  1970).

Jensen, Arthur R. "How Much Can We Boost IQ and Scholastic
  Achievement," *Harvard Educational Review*, 39 (Winter 1969).

Kallen, Horace. "Democracy Versus the Melting Pot," *Nation* (1915).

Kane, John J. "The Social Structure of American Catholics," *The
  American Catholic Sociological Review*, 16 (March 1955).

Karier, Clarence J. "Business Values and the Educational State," Urbana, Ill.: University of Illinois, 1971 (mimeographed).

———. "Elite Views on American Education," in Walter Laqueur
  and George L. Mosse (eds.), *Journal of Contemporary Education*,
  6, New York: Harper Torchbooks, 1970.

———. "Review of Patricia Albjerg Graham's *Progressive Education:
  From Arcady to Academe*," *Educational Theory*, 20 (Spring 1970).

———. "Testing for Order and Control in the Corporate State," Urbana, Ill.: University of Illinois Press, 1971 (mimeographed).

Katz, Michael B. "From Voluntarism to Bureaucracy in United States
  Education," 1970 (mimeographed).

Lanzer, William. "The Pattern of Urban Revolution in 1848," in Evelyn
  L. Acombe and Marvin L. Brown, Jr. (eds.), *French Society and
  Culture Since the Old Regime*, New York: Holt, Rinehart and
  Winston, 1966.

Larson, Karen. "Review of Oscar Handlin's *The Uprooted*," *American
  Historical Review* (April 1952).

Lean, Arthur E. "Review of *Public Education of the Future*," *History
  of Education Journal*, 6 (Fall 1954).

Machlup, Fritz. "Comments on Human and Social Benefits of Universal

University Attendance," Paper presented to the 54th Annual Meeting, American Council on Education, October 7, 1971.

Maller, J. B. "The Economic and Social Correlatives of School Progress in New York City," *Teachers College Record* (May 1933).

Mayer, Lawrence A. "Young America: By the Numbers," in "The Stairway of Education," *Fortune*, 77 (January 1968).

Mayo, M. J. "The Mental Capacity of the American Negro," *Archives of Psychology*, 28 (1913).

McCarthy, Colman. "40 Million Americans and a Broken Odyssey," *Washington Post*, July 13, 1970.

McNeill, H. *et al.* "Demographic Information by Health Area," New York: Maimonides Medical Center, Program Evaluation Section, 1967 (mimeographed).

Morison, Samuel Eliot. "Faith of a Historian," *American Historical Review*, 56 (January 1951).

Moynihan, Daniel P., *et al.* "The Schools in the City," *Harvard Today* (Autumn 1967).

National Council of Education. "Standards and Tests for Measuring the Efficiency of Schools or Systems of Schools," U.S. Bureau of Education Bulletin No. 13 (1913).

National Labor Committee. "Child Workers of the Nation," Washington, D.C., 1909.

National Society for the Study of Education. *The Twelfth Yearbook* (1913); *The Fifteenth Yearbook,* "Junior High School" (1916); *The Sixteenth Yearbook,* "Urban School Problems" (1917).

"The Negro American," *Daedalus*, 94 (Spring 1965) and 95 (Winter 1966).

Odum, H. W. "Negro Children in the Public Schools of Philadelphia," *The Annals of the American Academy of Political and Social Science*, 44 (September, 1913).

Palmer, L. E. "New York's Truancy Problem," *Charities*, 17 (February, 1906).

President's Commission on Education Beyond the High School. "Second Report to the President," Washington, D.C.: Government Printing Office, 1957.

Riis, Jacob A. "Playgrounds for City Schools," *Century*, 48 (1894).

Saloutos, Theodore. "Exodus—U.S.A.," in O. F. Ander (ed.), *In The Trek of the Immigrant*, Rock Island, Ill.: Augustana College Library, 1964.

Schrag, Peter. "The Decline of the WASP," *Harper's* (April 1970).

———. "Growing Up on Mechanic Street," in *Out of Place in America*, New York: Random House, 1971.

———. "Is Main Street Still There?" *Saturday Review*, 53 (January 17, 1970).

Schrank, Robert, and Skein, Susan. "Yearning, Learning, and Status," in Sar Levitan (ed.), *Blue Collar Blues*, New York: McGraw-Hill, 1971.

Shanker, Albert. "The Big Lie About the Public Schools," *The New York Times* (May 9, 1971), U.F.T. Column.

Shaw, A. H. "The True Character of New York Public Schools," *World's Work*, 7 (1903–1904).

Shils, Edward. "Plentitude and Scarcity: The Anatomy of an International Cultural Crisis," *Encounter* (May 1969).

Siegel, David. "Survey and Survey Trends," *Review of Educational Research*, 12 (December 1942).

Silberman, Charles. "The City and the Negro," *Fortune Magazine* (March 1962).

Sizer, Theodore. "The Schools in the City," *Harvard Today* (Autumn 1967).

Smith, Henry Lester, and O'Dell, Edgar Alvin. "A Bibliography of School Surveys and References on School Surveys," Vol. 8 (Indiana University, September-November 1931).

———. "A Supplement to a Bibliography of School Surveys and References on School Surveys," Vol. 14 (Indiana University, June 1938).

Smith, Timothy L. "Native Blacks and Foreign Whites: Varying Responses to Educational Opportunity in America, 1880–1950," paper presented at the Annual Meeting of the American Education Research Association, Spring 1970 (mimeographed).

Sommerskill, John. "Dropouts from College," in Nevitt Sanford, *The American College*, New York: Vintage, 1962.

Strayer, George D. "Progress in City School Administration During the Past 25 Years," in Alvine Eurich (ed.), *The Changing Educational World, 1905–1930*, Madison, Wisc.: University of Wisconsin Press, 1931.

———. "Retrospective and Prospective," New York: Institute for Educational Research, Teachers College, Columbia University, 1931.

Subcommittee of the Joint Legislative Committee to Investigate Procedures and Methods of Allocating State Monies for Public School Purposes and Subversive Activities. "Report," 1944.

Thernstrom, Stephen. "Poverty in Historical Perspectives," in Daniel P. Moynihan (ed.), *On Understanding Poverty*, New York: Basic Books, 1969.

Tyack, David B. "City Schools at the Turn of the Century: Centralization and Social Control," Stanford, Calif.: School of Education, Stanford University, 1969 (mimeographed).

———. "Forming the National Character: Paradox in Educational Thought of the Revolutionary Generation," *Harvard Educational Review*, 36 (Winter 1966).

———. "The Perils of Pluralism: The Background of the Pierce Case," *American Historical Review*, 74 (October 1968).

———. "Religious Folkways and the Law of the Land," in Paul Nash (ed.), *History and Education*, New York: Random House, 1970.

———. "This Period of Ferment May Be a Turning Point," *The New York Times* (January 11, 1971), Annual Education Review.

U.S. Bureau of the Census. Current Population Reports, Series P-20: No. 185 (July 11, 1969); No. 207, "Educational Attainment: March, 1970" (1971); No. 220, "Ethnic Origin and Educational Attainment, November, 1969" (1970).

———. "Historical Statistics of the United States, Colonial Times to 1957." (1960).

———. Supplementary Report, "Negro Population in Selected Places and Selected Counties," Series PC (S1), No. 2 (1971).

U.S. Department of Labor. "Annual Report of the Commissioner General of Immigration," Washington, D.C., 1921.

———. "The Condition of Negroes in Various Cities," 1897.

———. "The New York Puerto Rican: Patterns of Work Experience," Bureau of Labor Statistics Regional Report No. 19, May 1971.

———. "A Sharper Look at Unemployment in U.S. Cities and Slums," January 1967.

———. Urban Employment Survey—Report No. 1: "Poverty—The Broad Outline, Detroit" (February 1970); Report No. 2: "Poverty —The Broad Outline, Chicago" (March 1970).

U.S. Immigration Commission. "Reports," Senate Document No. 747 (1911).

Van Denburg, J. K. "Causes of the Elimination of Students in Public Secondary Schools of New York City," No. 47, "Contributions to Education" (New York: Teachers College, Columbia University, 1912).

Vecoli, Rudolph J. "Contadini in Chicago: A Critique of *The Uprooted*," *Journal of American History* (December 1965).

Violas, Paul. "Jane Addams and Social Control," Urbana, Ill.: University of Illinois, 1969 (mimeographed).

Woodward, C. Vann. "White Racism and Black 'Emancipation'," *New York Review of Books* (February 27, 1969).

Zuckerman, Michael. "Review of Lawrence Cremin's *American Education: The Colonial Experience, 1607–1787*," *American Association of University Professors Bulletin*, 57 (Spring 1971).

SCHOOL SURVEYS AND REPORTS

Allen, William H. "Cooperative and Constructive Survey—The Allen Survey: New York City Schools" and "School Survey Committee Report" (New York: New York City Board of Education, 1924 and 1925).

Biennial Survey of Education in the United States: Statistics of Public Schools; 1936–1937, 1938–1939, 1954–1955 (Washington, D.C.: U.S. Office of Education).

Birnbaum, Robert, and Goldman, Joseph. "The Graduates: A Follow-Up Study of New York City High School Graduates of 1970" (New York: City University of New York, 1971).

Boston Board of Education. "Survey of Boston Schools," 1924.

Chicago Board of Education. "Chicago Public Schools," 1924.

Committee on Schools, Fire, Police and Civil Service of the City Council of Chicago. "Recommendations for the Reorganization of the Public School System of the City of Chicago," 1916.

Connecticut State Board of Education. "Attendance and Child Labor," 1905.

──────. "A Study of the Costs of Secondary Education in Connecticut," 1924.

Detroit Board of Education. "Age, Grade, and Nationality Survey," Research Bulletin No. 3, 1920.

Educational Commission of Cleveland. "Report on Cleveland Schools," Cleveland Board of Education, 1906.

Ettinger, William L. "Survey of the Junior High Schools of the City of New York," New York City Board of Education, 1923.

Graves, F. P. "Report of a Study of New York City Schools," Albany, 1933.

Hanus, Paul M. "School Inquiry of New York Public Schools," New York: Board of Estimate and Apportionment, 1913.

──────. "Survey of New York City Schools, 1911," New York: New York City Board of Education, 1911.

Harper, William R. "Report of the Educational Commission of the City of Chicago," 1898.

Havighurst, Robert J. "The Public Schools of Chicago: A Survey Report," Chicago Board of Education, 1964.

Hull, Osman R. "Survey of the Schools of Los Angeles," 1934.

Irwin, Elizabeth. "Study of New York City Truants, 1915," quoted in Crime Commission of New York State, "Study of 201 Truants in New York City Schools," Commission on Causes and Effects of Crime, 1927.

Jensen, William. "Annual Report of the Superintendent of Schools: Our Public Schools, 1948–1950," *The New York Times* (January 19, 1950).

Laidlaw, Walter. "Population of the City of New York, 1890–1930," New York: New York Census Committee, 1932.

──────. *Statistical Sources for Demographic Studies, Greater New York, 1910*, New York: New York Federation of Churches, 1912.

──────. *Statistical Sources for Demographic Studies, Greater New York, 1920*, New York: New York Federation of Churches, 1912.

Mann, Horace. "Seventh Annual Report of the Board of Education" together with the "Seventh Annual Report of the Secretary of the Board," Boston, 1844.

Miles, H. E. "Industrial Education: Report of the Committee on Industrial Education," New York: National Association of Manufacturers, 20th Annual Convention, 1915.

Moehlman, Arthur B. "Public Education in Detroit," Detroit, 1925.

New York City Board of Education. "Annual Report of the Superintendent of Schools," 1900, 1901, 1902, 1903, 1904, 1905, 1906,

1907, 1908, 1909, 1911, 1913, 1915, 1934, 1937, 1948, 1957, 1966.

――――. "Fiftieth Annual Report of the Superintendent of Schools," 1948.

――――. "The Harlem Project: The Role of the School in Preventing and Correcting Maladjustment and Delinquency," 1949.

――――. "Special Report: Elementary Schools of New York City— Their Problems and the Efforts That Are Made to Solve Them," 1934.

――――. "Special Report: Handicapped and Underprivileged Children," 1934.

――――. "A Ten Year Report from Principals of Districts 23 and 24," 1937.

――――. "Youth in School and Industry," 1937.

――――. Commission on Integration. "Sub-Commission Reports," 1956.

――――. Committee on Maladjustment and Delinquency. "Report on Delinquency in New York City Schools," 1937.

New York City Commission on Human Rights. "Equal Employment Opportunity and the New York City Schools," January 25–29, 1971.

Pennsylvania Department of Public Instruction. "Report of the Survey of the Public Schools of Philadelphia," Public Education and Child Labor Association of Pennsylvania, 1922.

Ross, Frank. *School Attendance in 1920*, Washington, D.C.: Government Printing Office, 1924.

State Education Commissioner's Advisory Committee on Human Relations and Community Tensions. "Desegregating the Public Schools of New York City," New York: Institute of Urban Studies, Columbia University, 1964.

Strayer, George D. "Age and Census of Schools and Colleges: A Study of Retardation and Elimination," U.S. Bureau of Education Bulletin No. 5, 1911.

――――. "Foreword to the Fundamentals in Education: The Hartford Schools," New York: Teachers College, Columbia University, 1937.

――――. "Planning for School Surveys," New York: Institute for Educational Research, Teachers College, Columbia University, 1948.

――――. "Report of New York City Schools: Interim Report," 1943.

――――. "Report of the Survey of the Public School System of Baltimore, Maryland," New York: Institute for Educational Research, Teachers College, Columbia University, 1922.

――――. "Report of the Survey of the Public School System of Washington, D.C.," Subcommittee on D.C. Appropriations of the Senate and House of Representatives, 1949.

――――. "School Survey;" Paterson, New Jersey (1918); Baltimore, Maryland (1921); Lancaster, Pennsylvania (1924); Fort Lee, New Jersey (1927); Missouri (1939); Holyoke, Massachusetts (1930); Chicago (1932); Hartford, Connecticut (1937); New York City (1940); Boston (1944).

————. "The Survey Staff" in "School Survey: Lynn, Massachusetts," 1927. New York: Institute of Educational Research, Teachers College, Columbia University.

Superintendent of Schools, Archdiocese of New York. "Rise in Non-Public School Attendance in Manhattan and the Bronx," January, 1969 (Unpublished memorandum).

Van Sickle, James H. "Brookline, Massachusetts," 1917.

Wilson, F. M., and Krugman, M. "Studies of Student Personnel," Albany: Planning Committee of the Cooperative Study of Vocational Education in New York City Schools, 1951.

Works, George A. "Philadelphia Public School Survey," Philadelphia Board of Education, 1937.

NEWSPAPERS AND PERIODICALS

The Annals of the American Academy of Political and Social Science, 98 (November, 1921).

Charities (1899 through 1909).

Crisis (1910 through 1921).

Educational Review, 38 (1908).

Mental Hygiene, 7 (1923).

Milestones (1893 through 1898 and 1903 through 1910).

The New Republic (January 29, 1916), editorial.

The New York Times (for the years: 1930 through 1950, 1955 through 1959, 1970, 1971); Annual Education Review (January 11, 1971); New York Times Index.

Outlook (1906, 1914).

School and Society, 12 (1920).

School Journal, 80 (1913).

Survey (1909 through 1920).

Index